IMAGES OF WALES

CARDIFF PUBS
AND BREWERIES

IMAGES OF WALES

CARDIFF PUBS
AND BREWERIES

BRIAN GLOVER

The
History
Press

Frontispiece: Steward Peter Patten (left) watches as barmaid Ann
Roddies pours the final pint at the Irish Club in Herbert Street in
1968. It claimed to be the oldest club in Cardiff, and was originally
the Hastings Hotel.

First published 2005 by Tempus
Reprinted 2017 by
The History Press
The Mill, Brimscombe Port,
Stroud, Gloucestershire, GL5 2QG
www.thehistorypress.co.uk

British Library Cataloguing in Publication Data.
A catalogue record for this book is available from the British Library.

ISBN 978 0 7524 3110 9

Typesetting and origination by Tempus Publishing Limited.
Printed in Great Britain. by TJ International Ltd, Padstow, Cornwall

Contents

BUT I'VE LEARNT TO LOVE
MY ENEMIES!

On m'a appris à aimer mes ennemis.

Drinkers fight back against the stern
temperance preachers in this humorous
Edwardian postcard.

Acknowledgements

For providing many of the pictures in this book, I'd like to thank the *Western Mail & Echo* and Brain's Brewery in Cardiff, the former Whitbread Archive in London, the Bass Museum (now the Coors Visitor Centre) in Burton and the National Museum of Wales, plus Nigel Billingham and Stephen Jones for some of the Ely illustrations. I'd also like to thank Andrew Cunningham, John Hopkins and Keith Osborne for allowing me to use their extensive collections of brewing memorabilia. The staff at Cardiff Central Library and the Glamorgan Record Office also helped, as did the National Brewing Library at Oxford Brookes University and the Brewery History Society Archive in Birmingham. Finally, I'd like to thank the odd pint of Brain's Dark for refreshing the research.

one

THE BATTLEGROUND

In the middle of the First World War, another battle was raging. At its heart was alcohol. 'Drink is doing us more damage in the war than all the German submarines put together,' thundered Minister of Munitions, David Lloyd George, in 1915. 'We are fighting Germany, Austria and drink; and as far as I can see the greatest of these deadly foes is drink.'

The bitter struggle had already claimed a number of casualties. First to be shot down was all-day drinking in pubs. Under the stern gaze of DORA, the 1914 Defence of the Realm Act, a Central Control Board had been established in 1915 to imprison the liquor trade inside a barbed wire roll of regulations. As a strategic port, Cardiff was one of the first cities to be snared.

The new orders restricted weekday pub opening times to just five-and-a-half hours a day, from 12 p.m.-2.30 p.m. at lunchtime and from 6 p.m.-9 p.m. in the evening. There was no relaxation of the rules on Saturday. In fact, they were worse. Landlords had to throw the towels over the pumps an hour earlier at 8 p.m. The curtailed licensing hours initially applied only to areas deemed militarily important. This put Cardiff firmly in the firing line, with tight controls imposed in August 1915.

These were crushing changes. Though Wales was used to Sunday closing, the rest of the week customers had been able to drink from 5.30 a.m. until late at night. Harry Prickett of Cardiff Licensed Victuallers' Association (LVA) said the orders spelt 'absolute ruin' for many licensees. The president of the South Wales Brewers' Association, W.H. Brain, described them 'as a great blow to the trade'.

It was not the only blow. Another casualty was the beer itself. The brewing industry had to shoulder a huge burden of wartime taxation, doubling the cost of the cheapest pint from 2d to 4d between 1914 and 1916. Later it rose to 5d. At the same time, it had to struggle against a severe shortage of men and materials, and restrictions on output, which almost halved production from over 37 million barrels in 1913 to 19 million in 1917. The average strength of beer collapsed, from more than 5 per cent alcohol to below 3 per cent. The wishy-washy brew was unpopularly known as 'Government Ale' or 'War Beer'. Even treating or buying in rounds was prohibited.

But darker clouds were gathering over Cardiff. For the city was a central cog in the military machine. Through the port poured vast amounts of iron and coal from the South Wales Valleys. It was particularly important for steam coal, which powered the Royal Navy and the merchant ships which brought in vital food supplies. Cardiff was the global capital of coal, as crucial at the time for supplying the world's energy needs as the oil-rich states of the Middle East today.

The brewing industry had already been nationalised in one sensitive area. In Carlisle, where there were many important munitions works, the breweries and pubs were taken over by the state in June 1916, with many closed down in a determined attempt to stamp down on drunkenness among the workers. Cardiff could be next.

Lloyd George, who became Prime Minister in 1916, also had his eye on a higher, drier ambition. As one of two avowed teetotallers in the five-man war cabinet, he had supported ending the sale of alcohol in Wales since the 1880s. Now the prize seemed within his grasp. The United States was already moving towards total prohibition and expected its ally to follow. Food administrator Herbert Hoover issued veiled threats that an increase in American grain exports would be difficult to secure without an end to brewing in Britain. The choice seemed to be bread or beer.

One South Wales brewer, George Westlake, blamed the growing clamour on 'fanatical teetotallers' who were using the war to push their 'fiendish propaganda for the purpose of wiping out the trade'. He was right. Temperance leader, Sir Thomas Whittaker MP, told his followers that the conflict provided them with a great opportunity to strike 'while the overshadowing issues of the war are accustoming the people to restricted liberties'.

It was against this bleak background that leading Cardiff brewer, Brain's, issued a defiant notice in a local newspaper, the *Evening Express*, on Saturday 15 July 1916, warning people not to interfere in its business:

We believe there is to be an effort made by a few busybodies with time on their hands ... to promote the cause (if it be a cause) of prohibition: that is, they have the effrontery to suggest, because they have neither the head nor the stomach to appreciate good beer, therefore all men and women must be compelled to drink water. The suggestion is an insult to a hard-working, law-abiding, sober country, which has other matters on hand just now than answering silly questions of intemperate people who lack good taste as well as good manners. The British Empire – which they are oh so proud of! – was not built by water drinkers and conscientious objectors. Our fighting men and our working men and women have the liberty – and mean to preserve it – to decide whether they will drink water or merely wash in it. May we ask the clear-thinking and broad-minded folk of Cardiff to give a decided 'NO' if they are asked to sign any petition which suggests that the sale of beer, wines and spirits should be prohibited entirely because they are intoxicating when over-indulged in. Say 'NO' – then carefully shut the door.

This tirade shows just how much pressure the brewers in Cardiff were under. Their livelihoods were on the line. But it had long been a difficult trade in Wales, where the temperance movement had become such a powerful force in society since the 1830s.

Once beer was an essential part of the nation's diet. At a time when water was often contaminated – the diluted mud provided in Dowlais around 1810 was described as 'absolutely in motion with living organisms' – beer was the only safe drink. Even milk was suspect. Water used in brewing was boiled, while the added hops and alcohol created extra antiseptic powers. It was regarded as healthy and sustaining, being made from grain. Everyone drank ale, from small (weak) beer for breakfast to strong ale for special celebrations. It was seen as a minor miracle, a gift from God. No-one understood how yeast turned the sugary solution of malted barley into alcohol. They just knew it did. They called yeast 'Goddisgoode'.

Right: Leading Welsh politician Lloyd George was a powerful advocate of total prohibition, who rattled the brewing industry in Wales.

Below: Cardiff's vital role in supplying coal meant that docks pubs like the Old Sea Lock Hotel alongside the Glamorgan Canal were in the firing line in the First World War. Drinking was seen as a threat to the war effort.

One of Brain's beers today, The Rev. James, celebrates the fact that a minister, James Buckley, once owned a brewery.

The gospel and the glass had gone happily hand in hand for centuries. Monasteries were the earliest large-scale breweries. Church ales were specially brewed to help fund the upkeep of buildings, while many services, like funerals and weddings, traditionally featured a deluge of drink. Services were even held in pubs. Mass was celebrated in the Red Lion in Queen Street, before Catholic churches were built in Cardiff, until Bishop Baines banned the practice in 1835, saying it exposed his flock to liquid temptation. One of Brain's beers today, Reverend James, is a reminder that a Methodist preacher once owned a major brewery, Buckley's of Llanelli.

This cosy relationship changed in the second quarter of the nineteenth century. There were a number of reasons. First a safe, alternative drink had emerged out of the ships sailing into Cardiff – tea. From sales of 1 million lbs in 1730, it had shot up to 46 million lbs by 1847, as its price was brought within reach of the working class. Much more significantly, there had been a revolution, the Industrial Revolution. What was once a relatively stable, largely rural society, governed by local landowners and the Church, had been shaken to its foundations, as workers gathered together in large numbers around ironworks, mines and factories.

Across the Channel, there had already been a political revolution in France, followed by an extensive European war. The Wellington government, anxious to reverse a growing gin trade, fearful of unrest at home and keen to restore its popularity with the public, decided to roll out the tempting barrel of free trade in beer. Its Beer Act of 1830 was one of the most revolutionary social measures of the nineteenth century. It allowed any householder, on payment of a two-guinea licence, to sell beer. Previously, the number of licensed inns had been carefully controlled by local magistrates. Now it was open season. Almost anyone could run a pub. Within six months some 25,000 beerhouses had flung open their doors. The number of pubs in Britain nearly doubled in less than a decade.

Temperance Pledge.

📖   📖

I Promise, by GOD'S help, to abstain from all **INTOXICATING DRINK.**

Roath Road
Wesleyan Methodist Church,
Cardiff.

Signature Daisy Edna Holloway
W. E. Clogg 20/6/11

Nonconformist churches expected their members to sign the pledge, like this 1911 one from Roath Road Wesleyan Methodist church.

At the same time the duty on beer was abolished. Thus ale was both cheaper and much more widely available. The growing gangs of industrial workers, more affluent than their country cousins despite appalling working conditions, reached out eagerly for this fresh draught. Drunkenness spread, stumbling down the new, mean streets of back-to-back houses. The Church was appalled. The Revd Sydney Smith famously said, within a fortnight of the Act becoming law, 'Everyone is drunk. Those who are not singing are sprawling. The sovereign people are in a beastly state.'

Temperance preachers found a swelling audience. The first temperance society in Cardiff was set up in 1833, although initially it just promoted moderate drinking. But soon thousands signed teetotal pledges – in which people committed themselves to abstain from all intoxicating drinks – proudly hanging framed certificates on their walls. The movement was closely linked to the Nonconformist chapels. A Cardiff Total Abstinence Society was established in 1836, backed by local shipowner and Wesleyan Methodist, John Cory. Later the Cory Memorial Temperance Hall was built opposite Queen Street station, as a sturdy monument to his work.

A more extensive dry landmark was erected just a beer bottle's throw from the front of Brain's Brewery. 'Temperance Town' – street after street of houses and shops without pubs – was built on land reclaimed from the River Taff next to the main railway station between 1858 and 1864. Landowner Colonel Wood had stipulated that no 'tavern keeper, alehouse keeper or retailer of beer' be allowed. The main building was Wood Street Temperance Hall (later the Congregational church), which could seat 3,000 people. Next door was Raper's Temperance Hotel.

Many industrialists supported the movement. They were fed up with having to rely on workers who were unfit through drink or, more frequently, failed to turn up at all on Monday morning. John Guest of the Dowlais Iron Company ruled in 1831, after the Beer Act, that 'no

The Cory Brothers were the largest coal exporters in Cardiff and in 1896 John Cory built Cory Hall to support the temperance movement. The solid monument opposite Queen Street station was demolished in 1984.

A tram drives through Temperance Town in 1940, with Wood Street Congregational church on the left – and in the distance Brain's Brewery.

person employed in our service must have anything to do with keeping a public house or beer shop, and if any person now does so he must be warned to leave one of the two.' He employed solicitors to oppose new licences around his works and to close existing pubs.

Other industrialists, however, hedged their bets and profited from their workers' thirsty habits. The Rhymney Iron Company built its own brewery in 1839 to supply its men. Later this brewery was to play a major part in the brewing industry in the Welsh capital. Meanwhile the Melingriffith tinplate works in Cardiff rewarded its employees with beer notes, which could be exchanged for three to four quarts in local pubs. Some of these beerhouses were owned by company officials.

This ambivalent attitude to alcohol was reflected in the population at large. On the one hand they were told there was a devil lurking at the bottom of every glass. On the other many believed beer was highly nourishing, especially for manual labourers on limited diets. When one worker contemplated taking the pledge, his shocked friend said, 'Thee mustn't. Thee'll die.' Many stouts were sold as invalid stouts. The slogan 'Guinness is Good for You' persisted into the 1960s and the Irish brew was widely endorsed by doctors who prescribed it for convalescent patients. Ely Brewery in the 1950s still used the slogan 'For Strength and Stamina'. Wine was often sold as a tonic. A drop of spirits like brandy was used to revive the stricken. Even the very first 1832 teetotal pledge 'to abstain from all liquors of an intoxicating quality, whether ale, porter, wine or ardent spirits' carried the rider 'except as medicine'.

But the teetotal preachers were savage in their condemnation of the drinks industry. In a book aimed at children, *The Band of Hope Companion* of 1891, John Wesley was quoted condemning distillers: 'They murder His Majesty's subjects wholesale, neither does their eye pity or spare; they

Left: The South Wales Temperance and Band of Hope Union told school children, on this 1925 certificate awarded to a pupil at Lansdowne Road Boys' School, that 'Alcohol is the Great Ally of Every Disease'.

Below: Temperance campaigners pulled no punches in their attack on alcohol, as shown by this cartoon, 'The First Drop and the Last Drop'.

drive them to hell like sheep.' The close links with the chapels meant that teetotalism became deeply rooted in Nonconformist Wales, much more than in England. It became part of the country's religion that drunkenness was the basis of all evil, while pubs were dens of vice. It was there that 'the youth of our country are corrupted, honesty and virtue are bartered, and health and character are drowned within its fatal cup,' claimed the *Teetotal Times* of 1849. Wales was soon swimming in fervent sobriety.

'Hundreds have been misled by these cursed drinks from their profession, and have backslided across the alcoholic half-pints, and have been cast like Jonah, into the sea of intemperance,' wrote a contributor to one of the many Welsh temperance magazines in 1840 (there were 14 Welsh-language journals by 1850). 'But teetotalism like the whale swallows them and casts them on dry ground; blessed be God for such a glorious cause. Hallelujah!'

Such fervour could not be maintained. After seriously alarming the brewers, many let their teetotal pledges slip. Temperance gave way to temptation. But neither did the passion disappear. It kept coming back as wave after wave of religious and temperance revivals swept Wales. And with the chapels the driving force behind the movement, many teetotal campaigners soon came to focus on one highly sensitive issue – drinking on Sunday. Drunkenness was bad, but getting smashed on the Sabbath was sacrilegious. Bishop Wordsworth of Lincoln in 1870 damned those behind the bar. 'Publicans know that their souls are in peril so long as they continue to sell liquor on the Lord's Day.'

Sundays had always been seen as a special case, with shops shut and pubs supposed to close for divine service. But this vague requirement was widely ignored, until the Lord's Day Act of 1848 decreed that pubs should not open before 1 p.m. Most drinkers did not object too strenuously to this longer lie-in, but in 1854 an attempt to introduce afternoon closing from 2.30 p.m to 6 p.m, with evening closing at 10 p.m., sparked riots in London. A hurriedly amended Act in 1855 reduced Sunday afternoon closing to just two hours, from 3 p.m. to 5 p.m., with final drinks at 11 p.m.

However, in Nonconformist Wales, this was never regarded as enough. Already Calvinist Scotland, through the Forbes-Mackenzie Act of 1853, enjoyed complete Sunday closing, while the Irish Sunday Closing Act of 1878 shut all bars apart from those in major cities, which were allowed to open from 2 p.m. to 7 p.m. Wales's problem was that, as far as legislation was concerned, it was usually lumped together with England. And its debauched big neighbour wanted to prop up the bar on Sunday.

However, pressure was mounting in Wales. Petitions proliferated during the late 1860s and early 1870s. The issue became a moral crusade. The militant Grand Lodge of the Good Templars, meeting in Swansea in 1875, called for a distinct temperance Bill for Wales, banning Sunday pub opening. With the country's strong chapel culture, this proposal had widespread support. A canvass of nineteen towns between 1878 and 1880 claimed that 38,443 households were in favour with only 717 against.

The only significant opposition was in Cardiff. There the Licensed Victuallers' Association organised a petition against the Bill, which attracted nearly 17,000 signatures. This shows how distinctly different Cardiff was from the rest of the nation. It was a cosmopolitan port, with a large floating population of many races. More than 10 per cent of those who signed were seamen.

But the vast majority of Welsh MPs were in favour. The Bill was first introduced in 1879 by John Roberts, MP for Flint. Most of the opposition came from English MPs, worried that a successful measure would leak across the border. Blocked by pressure of other business, it was eventually forced through by Prime Minister Gladstone close to midnight on Saturday 20 August 1881. Battling to the last, Cardiff MP Hardinge Giffarde had tried unsuccessfully to gain exemption for his constituency on the grounds of its large, migratory population.

The brewing industry was furious, fearing 'prohibition by degrees'. Cardiff's licensees counted the cost. Sunday was their busiest day, when workers had most money in their pockets and most time on their hands. But the Welsh temperance movement was jubilant. 'Cardiff is like Heaven upon Earth on Sundays since the passing of the Act,' said Alderman Sanders.

Another former mayor, Alderman Trounce, in a book, *The Case for Sunday Closing*, published in 1899, recalled:

> For several years prior to the passing of the measure, my residence was at the Bute Docks, from which district there is but one main thoroughfare to the town and to various places of worship – Bute Street. This street was at that time on Sundays too often a reproach to civilisation, the scene of drunken brawls and rowdyism. With public houses at every corner, frequented by sailors and others of all nationalities, respectable pedestrians were frequently subjected to gross insults, and the sound of most offensive language. I have visited Bute Street several times since the Act in question became law, but the scene has changed – order and quiet prevail where once there was so much disorder and riot. Our police have easy times compared with their onerous and trying duties of twenty years ago.

However, behind the doors off Bute Street, another, darker world of illicit drinking dens lay hidden in the shadows – the shebeens. Like prohibition in the United States in the 1920s, the Sunday Closing Act bred a criminal culture which was supported by many members of the public. Drinkers felt the deprivation more keenly in Wales, since the national drink was beer, which was difficult to keep in condition and awkward to carry around before mass-produced bottles appeared. In Scotland and Ireland, a hip-flask of whisky could more readily relieve the pain of a shut bar.

David Davies, assistant editor of the Cardiff morning newspaper, the *Western Mail*, which campaigned against the Act, investigated the extent of shebeens in 1889. Using information from the police and aggrieved publicans, he claimed there were 480 in Cardiff, concentrated in the centre, including, ironically, Temperance Town (137), Roath (91), Cathays (79), Docks (57), Newtown (45), Grangetown (40) and Canton (38). Many were run by labourers, particularly Irish navvies.

The Tory newspaper asserted that the Liberal Act had increased drunkenness and disorder, also highlighting the problem of the bona fide traveller, who under the Act was allowed refreshment on Sunday. This clause was widely abused, with crowds pouring out of Cardiff for a drink in seaside Penarth or country villages. 'Field clubs' multiplied in summer, while membership of established clubs, which avoided the licensing restrictions, surged by 3,000 in Cardiff in the twelve months from June 1882, an increase of 900 per cent over the previous year. Others simply crossed the nearby border into Monmouthshire, which was not included in the legislation until 1921.

Fifty years later, in 1933, J.C. Griffith-Jones wrote colourfully in the *South Wales Echo*:

> The scenes were indescribable. When time was called on Sunday nights, the highway between Newport and Cardiff was turned into a Bedlam. A vast army, more merry than bright, rent the night with ribald songs and rolled all the way home. Many slept off the effects of these orgies in ditches or awoke to a hopeless dawn in a police cell. In Cardiff there were fierce Sunday night clashes between the rebels and the police in Mary Ann Street, Gough Street and Adelaide Street. Pokers as well as invective were in evidence.

When Lord Aberdare, who had piloted the Bill through the House of Lords, admitted that he could no longer support the Act, given the fresh evidence from David Davies, a Royal Commission was set up in 1889 to look into its effects. The temperance movement was appalled,

HERE'S TO THE GOOD OLD BEER DRINK IT DOWN

An Edwardian postcard of two red-nosed drinkers reflects the temperance view that drinking was all about getting drunk and losing respect.

particularly since the five-man panel contained four Tories. This gave the brewers their chance to break free from the Sunday straitjacket.

County councillor Samuel Brain said the Act 'certainly drove the people into illicit drinking'. He told the Commission:

> Since the Act came into operation in 1882 nearly all the breweries within a radius of 100 miles of Cardiff have sent local agents here, and local men have had to go in very largely for the small cask trade. The great majority of the private houses in certain districts here have in immense quantities of this beer on Saturday nights, and in fact right up to one or two o'clock in the morning. It was nothing unusual for the well-conducted licensed victualler to stand at his door on Saturday nights, after he had closed his premises at 11 o'clock, and see brewers' drays and agents' carts delivering seventy or eighty small casks into private houses within 100 yards of his premises.

Mr Lloyd, a salesman for William Hancock in Cardiff, showed the extent of this trade, largely conducted through a new breed of street-corner wholesaler. Sales of small four-and-a-half and nine-gallon casks from the company's three breweries in the town (the Bute Dock, Phoenix and South Wales) had shot up from 872 in 1880-81, before the Act, to 39,055 in 1888-89, the bulk being in pins (nine gallons). Hancock's own pubs took much larger barrels or hogsheads.

Frank Granger of the Bedford Brewery, in the suburb of Roath, when asked who was using these small casks, replied, 'In many cases people who are living in almost shanties, will have four

17

Artist John Hassall, who in 1909 designed the famous 'Skegness is so Bracing' poster showing a jolly fisherman skipping along the beach, joined the fight against Sunday closing – but with no success in Wales.

or five or six, and I have known even seven and nine a week. They nearly always have it in on Friday morning or Saturday, and as a rule every cask is empty on Monday morning, and we collect the empties then.' He added, 'I know one or two who do not work at all, but simply live on the proceeds of shebeening.'

While the Royal Commission was taking its evidence in Cardiff in October 1889, 2,000 men marched through the town demanding 'No Compulsory Sunday Closing'. An election in the West Ward next month was dominated by the issue. The Tory E.J. Smith, campaigning for 'The Liberty of the Working Man', won with a substantial majority. He was the president of Cardiff LVA.

The drinks trade must have felt it was a good sign. But the view from the streets of Cardiff was not the view from the rest of Wales. The temperance lobby had also presented many well-briefed witnesses. When the Commission reported, its conclusions must have been hard for the brewers to stomach. It refused to consider either the repeal of the Sunday Closing Act or a modification to allow some limited pub opening hours, and even rejected sales off the premises. It ruled:

> We are convinced that a change in this direction would be so unwelcome to so vast a majority of the population in so large an area of the Principality, that we do not think it ought to be forced on this large area for the sake of a possible benefit to the rest of the country. Moreover, we find an almost complete absence of evidence of a desire for such an amendment on the part of those classes who would be most likely to require or use it.

In other words, though the drinks industry had spoken up, precious few of its customers had come forward to complain. Outside of Cardiff, a few other parts of Glamorgan and a limited area around the brewing centre of Wrexham, Wales solidly supported Sunday closing. The Commission had not swallowed the scare stories of widespread disorder, even if there were local difficulties in Cardiff. It was a sobering prospect for the brewers. For they knew it meant they were staring down the dark tunnel of total prohibition.

The Commission even sought to tighten the Act. It tried to knock bogus travellers off the road by requiring licensees to take their names and addresses for police inspection. Police powers against shebeens were increased and clubs more closely controlled. W.R. Lambert in his book *Drink and Sobriety in Victorian Wales* concluded that the improved Act increased sobriety in Wales 'with the notable exceptions of Glamorgan in general and Cardiff in particular, where the alcohol dripped through the dotted line'. However, he believed 'the result was to bring the law into disrepute', as publicans and drinkers sought to evade the Act, while the police 'were forced to engage in ludicrous forms of espionage in order to discover illicit drinking practices'.

The teetotallers had no such reservations. They were triumphant. Leading advocate Arnold Hills exulted, 'The trumpets are sounding for the Armageddon of Alcohol ... We stand upon the threshold of great events, and as far as the Dominion of Drink is concerned, we are privileged to see the beginning of the end. Already the chains are clanking in the Pit of Prohibition, where the devil of strong drink shall be bound for full a thousand years.'

The movement now expected the dry tide to roll over the rest of the week in Wales. Pubs would be shut for good and breweries closed down. Lloyd George claimed in 1891 that three-quarters of Welsh people were in favour of complete prohibition. The first Local Option Bill for Wales had been introduced into the Commons in 1887. This allowed areas to vote to ban alcohol or at least reduce the

The brewing industry mounted a vigorous leaflet campaign against local option, arguing it would lead to prohibition and loss of liberty.

number of pubs. Delayed at first, it received a major boost with the Royal Commission's favourable report on the Sunday Closing Act. A third Bill saw its second reading carried by six votes in 1891. Lloyd George boasted about 'our splendid victory'. The *Alliance News* trumpeted, 'The outpost has fallen and the capture of the main citadel is but a question of time.' They crowed too soon.

The brewers and distillers were determined not to allow the Bill to succeed. Once local votes went ahead in Wales, pressure would mount to allow them in England. Prohibition would creep across the land. The Marquis of Carmarthen, a director of Hollands Gin, moved sweeping amendments in committee. Held up at every stage, the Bill was finally dropped in July. Another met the same fate in 1893. One supporter, Herbert Lewis of Flintshire, claimed 'every cubic inch of the House of Commons is charged with obstruction'.

Lloyd George decided to give up banging his head against the brick wall of Westminster. He decided that home rule would have to come first, and then prohibition would follow once the country had escaped 'the brewers' ring which seems to govern England'. But he was to gain revenge. First, as Chancellor of the Exchequer in 1909, he imposed a huge additional taxation burden of £4 million on the industry by increasing duties on breweries and pubs. And then in the First World War he moved in for the kill, as fighting raged across Europe.

With many countries joining the United States in embracing prohibition, the brewing trade was in every sense drinking in the last-chance saloon. As a first step, Lloyd George favoured the nationalisation of the whole brewing industry, which was seriously considered in 1915 and again in 1917. But the cost was far too great. However, it was dangerous to defy a powerful ally. In March 1917, Food Controller Lord Devonport moved to meet American demands, by ordering that brewing be limited to 10 million barrels, just 28 per cent of pre-war production.

This target was never reached. Ministers came to realise they had squeezed the public's pint too far. The Home Secretary, Sir George Cave, told the House of Commons in July 1917, 'The beer shortage is causing considerable unrest, and is interfering with the output of munitions and with the position of the country in this war. There is unrest, discontent, loss of time, loss of work and in some cases strikes are threatened and indeed caused by the very fact that there is a shortage of beer.'

Beer was no longer a problem. Shortage of beer was. Restrictions were relaxed and output rose to 23 million barrels in 1918. But the Great War had given the brewers a great fright. It wasn't so much the harsh regulations which were alarming – these were to be expected during an emergency – as the atmosphere of official hostility. Beer had been branded a danger, an intoxicant which could undermine the war effort. The brewers were in the opposite trench along with the enemy.

And if they had hoped for a restoration of the merry pre-war days once the conflict was ended, they were to be bitterly disappointed. There was to be no return to the cheap and staggeringly strong ales of the Edwardian era. The huge weight of wartime taxation remained in place, keeping the price of beer up and its strength down. There was also no return to all-day drinking. When the Central Control Board was abolished in 1921, many of its restrictions were continued in the 1921 Licensing Act, notably the limitations on pub hours. Bars continued to be shut for most of the morning and close in the afternoon, with last orders in the evening at 10.30 p.m. or earlier.

The brewing industry had been let out on bail under strict conditions. It knew that trouble in the streets or another emergency – and within little more than a decade another war was looming on the horizon – could mean it would be back in jail, facing a possible death sentence. And there were many in Wales pushing for the ultimate penalty. The teetotal movement was still a formidable force. There were a host of organisations under the temperance flag. The Cardiff trade directory listed eight, including the Good Templars, the Rechabites, the South Wales Auxiliary, the Sons of Temperance and the Women's Temperance Association.

TEN NIGHTS IN A BAR-ROOM.
A TEMPERANCE DRAMA, IN FIVE ACTS.
BY WILLIAM W. PRATT.
First Performed at the National Theatre, New York, September, 1858.

The temperance movement not only produced novels supporting their cause, but also took to the stage with plays like *Ten Nights in a Bar-Room.*

Their strength of feeling is illustrated through the novels produced by the Temperance Publishing Company. *The Tavern Across the Street* by Derwenydd Morgan was published in 1915, 'to encourage the great army of temperance workers, who are today fighting this twentieth-century enemy – the curse of drink in Wales'. The story concerned the downfall of a farmer, Sam Rosser, who came to the imaginary mining town of Cambria to run a pub. Soon he was sliding into hell. 'The drinking, the swearing and the fighting on many a Saturday night in the Fox and Geese were enough to make humanity shudder.' The landlord's daughter Matilda killed herself after being wooed and abused by the aptly named, heavy-drinking Squire Hancock.

The temperance campaigners, led by a reformed drinker, won the town elections. The Fox and Geese declined. 'The people who called at the tavern now were those who had reached the lowest strata of society – the station known the world over as the gutter – the last stage of the drunkard, men who had nothing more to lose for their all had been sacrificed on the altar of strong drink. An ogling, drinking, thieving class, who ought to have died long ago, when all

"The Long Pull."

Copyright

The temperance campaigners' hatred of the pub is grimly illustrated in this Victorian print, showing death behind the blind publican drawing drinkers towards an early grave.

that was manly in them died.' The pub shut and was transformed into a temperance hotel. The final chapter was devoted to a meeting to form a Welsh Prohibition Party to end 'the accursed liquor trade'. It concluded, 'It will be a strenuous fight and a bitter one, but victory will be ours. Ere many years roll by ... every tavern will be closed. The white flag of Prohibition will wave on every hill, and our children growing up in Temperance Wales will wonder that we tolerated such an institution as an open tavern in our land.'

It was in the teeth of such fanatical opposition that brewers in Wales carried out their trade. It was never as comfortable a business as in England. Many in Cardiff still feared they could be swept away by the waves of teetotal beliefs. The sinful city was a pocket of pleasure in a stern, dry nation. It could be mopped up at any time.

two

THE BREWERIES

One of the least-celebrated facts about Cardiff is that it was once a major brewing centre. Beer in the capital was much more than just Brain's. It also had a substantial body. At its peak there were as many breweries as Wrexham, almost twenty steaming away to slake the thirsts of the dockers, sailors and railwaymen delivering ton after ton of coal to the world.

But unlike Wrexham, which was known as the 'Burton of Wales', the brewing industry never dominated Cardiff. It was seen as just another service trade, helping to lubricate the flow of the all-important black gold. Brewing was often ignored in the trade profiles of the town or noted at the tail-end, along with pickle manufacturers and dog biscuit makers.

Another often-forgotten fact, as 2005 celebrates both Cardiff's 100th anniversary as a city and its fiftieth anniversary as the capital of Wales, is that the settlement was little more than a curious dot on the map until the nineteenth century. It had a long but far from prosperous history. It was just a huddle of houses and inns around a decaying castle, an old but rather rotten borough. In the first official census of 1801 it recorded a population of 1,870, which meant it had grown little, if at all, since it was a Roman fort and then a Norman seat of power.

Its chief strategic importance was as the lowest crossing point on the River Taff for one of the main east-west roads. But its position next to the turbulent waters of the Taff meant dwellings were subject to repeated flooding. The streets were mired in mud. The sea was also a threat to the low-lying plain on which it stood. A tsunami in 1607, which hit the Bristol Channel, washed away most of St Mary's church and set back the local economy for decades.

Even when the Industrial Revolution swept into South Wales, Cardiff at first was little more than a stagnant backwater. The main development was further west, where the coalfield met the sea at Swansea and Neath, and to the north in Merthyr Tydfil, where the blast furnaces were forging a new Iron Age. And it was the Merthyr ironmasters who triggered the expansion of Cardiff.

To get their heavy iron bars to customers, they needed better communications than packhorses and lumbering wagons. First they promoted the Glamorganshire Canal, which linked Merthyr to the sea at Cardiff in 1798. Cardiff's population crept up to 6,187 by 1831 as a result. But the real

Above: Low-lying Cardiff was always vulnerable to the sea, as starkly illustrated by this picture of the Red House in Ferry Road in 1977. The isolated pub even carried a naval navigation lamp. Sadly it was demolished in 2005 to make way for more waterfront apartments, despite a determined battle to save it.

Left: Cardiff was once packed with home-brew pubs. The last was the Marchioness of Bute in Frederick Street, run by Thomas Jenkins and his sons until the early 1920s, brewing 'celebrated home-brewed ales and stout' as well as bottling Bass and Guinness. It became a Brain's house.

explosion came with the opening of the Bute West Dock in 1839 and the Taff Vale Railway to Merthyr two years later, bringing a surge of coal and iron flowing through the ancient borough. Extra docks were built in 1855, 1864 and 1875 to keep up with demand. The population rocketed to 41,400 in 1861, 82,700 in 1881 and 164,300 by 1901. Cardiff was the boom town of the Victorian era. And the swelling crowds of workers brought with them terrible thirsts.

The first Cardiff trade directory, *Bird's* of 1796, reveals a small market town. There were no commercial brewers, just eighteen pubs, brewing their own beer. The main business in the drinks trade was producing and selling the raw ingredients. There were six maltsters and hop factors, although four were part-time. Business was not that busy. Elizabeth Jones was also a draper and hat dealer.

Many breweries sprang up in the nineteenth century. Some only lasted a few years. Henry Anthony ran the Castle Brewery in Frederick Street from 1862 to 1895; William Steeds was brewing at the Albion Brewery in Bute Street in the early 1890s. It had been founded by Thomas Williams around 1850 and was taken over by Miller's of Bristol in 1897. The Walpole Brewery in King's Road was formed in 1900 and went into liquidation in 1906.

The first common brewer, James Walters, mentioned in *Ridd's Directory* of 1813, was an extension of a malting business. His address in 1822 was St Mary Street, and he was brewing on the site of what later became Brain's Old Brewery. By 1829 Edward Thomas Bridgen Carter had taken over and the established business had a rival. John Wood's map of Cardiff in 1830 shows another brewery further up St Mary Street, opposite the Market House, run by William Williams.

But the Old Brewery was clearly the prime site with good well water, despite being perilously close to a bulging bend in the Taff. As the town surged ahead, so the brewery changed hands at a rapid rate. In 1835 it was being run by Watson and Phillips, in 1840 by Phillips and Andrews and by 1844 by William and Charles Andrews, the latter for the first time using the title the Old Brewery.

There was also now a third commercial competitor. John Thomas opened the New Brewery (later the Castle Brewery) in Great Frederick Street by 1840. Within two years this venture was in the hands of Elizabeth Thomas. But both rivals had an envious eye on the Old Brewery. By 1848 William Williams had taken over, abandoning his own brewery.

It was from this decade that Cardiff really catapulted ahead. And the population explosion was matched by a brewing boom. *Hunt's Directory* of 1848 lists six wholesale brewers, including William Nell's Eagle Brewery in St John Street, George Watson in the Cambrian Brewery off the High Street, and William Phillips in the Ship Brewery, Millicent Street. By 1850 there were eight, with John Reed opening the Bute Brewery in Whitmore Lane and Thomas Williams brewing in Bute Street. The glass of Cardiff beer kept frothing up. *Ewen's Directory* of 1855 records eleven breweries, including North and Low's Bute Dock Brewery, which was later to become the first building block of Hancock's brewing empire in South Wales. By 1885 the number had reached the heady height of eighteen.

Meanwhile, the highly prized Old Brewery continued to change hands. The new owner reflected the rapidly changing character of Cardiff. Frederick Prosser, who had taken over by 1858, was a young, well-travelled merchant, involved in shipping. While in India he had married the daughter of an Army officer, Maria Adams, and their daughter Adeline was born at the Old Brewery on September 19 1859. But his career was cruelly cut short when he died the following year while on a ship to India.

The tragedy gave the Thomas family their chance. They had established the New Brewery in Great Frederick Street, but preferred to sell up and buy the Old Brewery in 1862. John and Edward Thomas traded as Thomas Brothers, developing a small chain of 'retail establishments', including the famous Golden Cross. But it was their sister, Frances Elizabeth, who made the most significant connection. In 1872 she married Samuel Arthur Brain, who was to brand his name and initials permanently onto the rolling barrel of Cardiff beer.

And his arrival at the Old Brewery was helped by the temperance movement. For John Thomas, on hearing the bitter news of the success of the Welsh Sunday Closing Act in 1881, is said to have thrown down his pen and exclaimed, 'Anarchy'. He resolved to quit the troubled trade and his brother-in-law was well qualified to take over. Samuel Brain was a trained brewer, who had come to Cardiff in 1863 to work at and later manage the Phoenix Brewery in Working Street. He also had access to substantial capital. He came from a prominent Gloucestershire family and his uncle, Joseph Benjamin Brain of Bristol, was chairman of the West of England Bank. The two bought the Old Brewery in 1882 and ushered in a new era in the city's liquid history.

Brain's

Samuel Arthur Brain was a classic Victorian entrepreneur. Though he had bought the leading brewery in town, it was still little more than a three-storey stone building behind the Albert pub. In fact, all of Cardiff's many breweries were modest affairs. They struggled to keep up with the boom in demand, leaving the way open for larger English, Irish and even Scottish brewers to flood the town with their beers. Burton brewers like Bass and Worthington were particularly prominent, their casks being crammed into every available railway arch around the stations.

Samuel Brain was determined to change all that. He aimed to compete with the Burton brewers head on. But first he needed a much larger brewery. In 1885 he bought Watson's Cambrian Brewery in Womanby Street, but this was no answer and soon ceased brewing, the site being used for stores. It was when he was able to buy land adjoining the Old Brewery, that he was able to realise his ambition in 1887.

His new 'Old Brewery' was built just in time to be featured in Alfred Barnard's monumental work, *Noted Breweries of Great Britain and Ireland* (published 1889-1891). It was the only Cardiff brewery to be included in the four volumes, and the eager author was mightily impressed:

> As the visitor enters beneath the archway, he is confronted by the new brewhouse, which is an elegant structure built on the tower principle, with red and white bricks. It is almost fire-proof, iron, brick and concrete forming the principal elements in its construction, while its walls are nearly two feet thick. The iron joists which support the massive floors are as broad as a man's body and, with the numerous metal columns (seventy in number) which support them, weigh nearly fifty tons.
>
> This brewhouse, which is splendidly arranged, contains a costly and complete plant, said to be the largest in South Wales and, together with the new fermenting room, is fitted and ventilated on the most modern principle.

No expense had been spared. There were three steam lifts, while a new 400ft artesian well had been sunk. Two 150hp steam boilers of gleaming steel – 'the finest in Cardiff' – had been supplied by the local Mount Stuart Dry Dock Company.

After taking over the Old Brewery, Samuel Brain bought Dominick Watson's Cambrian Brewery in 1885.

Barnard was captivated. 'The mash tun is one of the handsomest we have ever seen, being constructed of cast iron, encased in varnished pine, and bound with massive brass hoops.' He declared the copper house 'the prettiest and most complete we have visited' with 'handsome iron palisading'. While the new fermenting department, with its white cedar tuns, was 'a delightful room, lighted by twenty-five windows'.

Samuel Brain was not so concerned with the 'polished brass rails', which dazzled Barnard, as the vastly increased capacity. The two new domed coppers each held 150 barrels. The new brewery could produce seven times more beer than the old one. The greatly enlarged size of the operation is best illustrated by the cellars, which had been extended beneath the yard and were linked to ground level by a steam-powered platform elevator made by Thomas & Sons of Cardiff.

'The cellars are all floored with vitrified panel bricks, which make a splendid firm pavement, and they are lighted throughout with gas,' wrote Barnard. 'The roofs of the cellars, having to support the heavy traffic in the yard, are unusually massive, and are concreted to iron girders laid on numerous iron columns ... Of the three, the central cellar is by far the most important, it being 300 feet in length, 200 feet in breadth, and containing nearly 5,000 casks.'

But this huge investment of £50,000 was still not enough to keep up with demand. Samuel Brain also had to keep the old brewhouse in operation alongside its towering big brother. Head brewer, Mr G.J. Gard, said that once 100 barrels a week was the average output, now it was more than ten times that, at 1,100 barrels. The plant was in use six days a week, except Sunday. 'No brewery in the kingdom has increased its output as rapidly as this,' he claimed. By 1896 it was up to 1,400 barrels a week. Even the new cellars could not cope, as Barnard discovered. 'As we passed through, the floor was covered with many hundreds of casks that had recently been filled, and we had some difficulty in picking our way between them.'

Boiling coppers at the Old Brewery, 1890

Barnard was captivated by the 'handsome iron palisading' in Brain's copper house. This sketch is from his book.

The Old Brewery cellars, 1890

The extensive new cellars in 1890.

A portrait of Samuel Brain looms above his great grandson Bill Rhys, who was chairman of Brain's from 1971 to 1985.

The new brewery also could not contain the restless energy of Samuel Brain. He was one of the founders in 1886 of the Cardiff Malting Company in East Moors, close to the docks. These extensive six-storey buildings supplied not only local brewers but also companies as far afield as Burton and Ireland. And he was chairman of Stevens & Sons, said to be the largest firm of wine and spirits merchants in South Wales.

But his hand reached far beyond business matters. He was elected to Cardiff Council in 1885, representing Grangetown, and also headed the poll for the Cardiff School Board in 1893. He was involved in many municipal committees and friendly societies, which often met in Brain's pubs – the family firm owned or leased over eighty by the end of the century. He was so prominent in the town's affairs that he was made Mayor of Cardiff in 1899. When Mafeking was relieved in 1900 during the Boer War, he toured the streets in an open landau. If the new brewery was a major landmark for the brewing trade in Cardiff, the man behind it, Samuel Arthur Brain, had become a local legend. He was one of the driving forces behind the development of the town.

In 1897 he turned the firm into a limited company, with a share capital of £350,000. This ensured it could compete in the race to buy up more pubs. But the brewery was still firmly controlled by the family, the board of directors including Joseph Brain's two sons, Joseph Hugh and William Henry Brain. Samuel Brain was chairman. He died in 1903, but his initials, SA, still linger on the lips as the name of Brain's premium bitter.

His partner, Joseph Brain, took over as chairman for four years, followed by his two sons. Demand for the company's ales continued to grow, particularly in the new market for bottled beers. The cramped central site could no longer keep up. An off-licence and bottling store had been opened as early as 1897, out in the suburbs in Nora Street, Roath. As this trade expanded, so a row of cottages was bought up and a larger site between Nora Street and Helen Street

29

Put Up To Be Knocked Down.

Samuel Brain was also an officer in the Artillery Volunteers and this was used by cartoonists to depict his many political battles on Cardiff Council.

Brain's soon tested motorised vehicles for deliveries, like this steam wagon driven by Fred Evans in 1907.

earmarked for a new brewery. The red brick tower was topped out in 1914, but the outbreak of war meant the first brew was not put through until five years later.

It was not an easy birth. By then the brewing industry was in turmoil from restricted licensing hours, heavy taxation and the threat of prohibition. The new brewery's site was also a red rag to a raging bull for the temperance movement, as it was built next door to Diamond Street United Methodist Chapel, which had been established with the help of the Cory family. At the same time, the original three-storey brewery behind the Albert was demolished in 1919. As the buildings were pulled apart, many wondered if the whole industry would suffer the same fate.

But the new licensing laws and heavy taxation on beer proved a blessing as well as a burden. With no return to all-day drinking and no return of the powerful, cheap brews of the past, the drunk, such a familiar feature of Victorian and Edwardian Cardiff, staggered off the streets. The legless table was on the slide. By 1932 the number of drunkenness cases in the city was 158. At its peak, in 1897, there had been 1,667 convictions. It was a major transformation.

By the 1930s Brain's were using more conventional vehicles, here being loaded in the brewery yard.

Local veto polls, introduced in Scotland in 1920, failed to catch on after initially forty areas (out of 900) voted for no licensed premises. Despite agitation, the polls were not extended to Wales. Heavy drinking was no longer seen as a major problem, while prohibition was becoming a dirty word. The complete banning of booze in the United States in the 1920s degenerated into a criminal culture of smugglers and illegal drinking dens. Instead of portraying a sober, upright, moral nation, the new image of America was one of gangsters and guns. The new face was not Uncle Sam but Al Capone. By 1933 the discredited law had been repealed.

With the dry smoke from the fire and brimstone preachers clearing away, Brain's was able to enjoy a more comfortable business, particularly once the Depression lifted. It was helped by the industry's first generic advertising campaign under the simple slogan, 'Beer is Best', launched in 1933. The Temperance Alliance responded with an adaptation of their own – 'Beer is Best Left Alone' – but it failed to dampen the brewing revival.

The teetotallers launched a final attack on alcohol when the Second World War broke out. They hoped that in the new emergency, the Government would again feel it necessary to imprison the industry. But attitudes had changed. Beer was no longer seen as a threat to the war effort. Instead, it was regarded as vital for maintaining morale in a crisis, and production was not only maintained but increased.

After the war, Brain's continued to build up its dominant position in Cardiff. Its name seemed to hang from every bridge and inn sign, while more pubs were added to its estate, bringing the total to around 120 houses. The largest single purchase was in 1956 of the five pubs belonging to Cardiff wine merchants, Greenwood and Brown. They included the city's best-known 'chop house', the Model Inn in Quay Street, the York Hotel on Canal Wharf and the Vulcan in Adam Street.

Few breweries have been so inextricably linked with one town. While rivals like Hancock's, Ely and Rhymney expanded across South Wales, Brain's seemed content to sit in the city. It barely expanded up the mining valleys, where its rivals ruled, only venturing out west into the Vale of Glamorgan. Part of the problem was that it was already brewing to near capacity.

While the New Brewery in Roath produced bottled and later keg beers and soft drinks, the Old Brewery continued to roll out its traditional draught ales into the city centre traffic, the drays rumbling out of an arch into busy St Mary Street. It looked increasingly out of place, an industry among shops and offices. It also faced a new threat. From the early 1960s, major combines were appearing, taking over local breweries to build huge brewing groups across Britain. But such was the affection for the company, the family were loath to lose their independence or move from their place in the heart and hearts of the capital.

Instead desperate measures were taken. The brewers were ejected from their office to make way for an extra fermenting vessel. But emergency moves like this could only temporarily solve the need to brew more cask ale, a market stimulated from the early 1970s by a vocal consumer movement, CAMRA, the Campaign for Real Ale. Fortunately, the market for bottled beer had declined and so some production could be moved to the New Brewery, though the bulk beer then had to be tankered to the Old Brewery for racking into casks. But soon the New Brewery was brewing to capacity as well. Brain's breweries were bulging at the seams.

In 1976, it was decided to expand the brewing capacity at the Old Brewery by 50 per cent. This was easier said than done, as the whole cramped site had to be reorganised and partly rebuilt while continuing production. The exercise in claustrophobic engineering was to take seven long years. At times the brewers thought the builders and contractors would never leave.

Being in the city centre, the new fermenting block alongside Caroline Street could not look like a crude industrial building. So it was given a glazed exterior, to appear like a smart office block. The old 140ft brick chimney, built in 1934, was replaced by a thin silver stack. Every yard of space was used. Rails were put round the flat roofs of the new buildings so they could store casks. In the brewhouse, two new mash tuns, two coppers and other vessels were installed. It was a sign of the changing times that the neighbouring Tabernacl Welsh Baptist church allowed access through its yard for ten months for this work to be carried out. The teetotal preachers must have been spinning in their graves.

Opposite: Brian Dean pours hops into the one of the new steel coppers.

Right: Normal service is resumed in the brewery yard in the 1980s, beneath the famous clock, after all the disruption.

By December 1979, the essential construction was complete and the equipment came on stream, capable of producing an extra 18 million pints a year. But the brewers could not wave goodbye to the builders. They then ripped up the brewery yard to renew the cellars beneath. This caused fresh problems. Drivers had to carefully steer round the gaping holes. Only in 1982 was the work finally concluded and the contractors evicted.

The £2.8 million expansion was a statement of intent. While all its local rivals had been taken over by the circling English brewing groups, Brain's had demonstrated it had faith in its future. The work had been completed just in time to celebrate the centenary of Samuel Arthur Brain's purchase of the Old Brewery in 1882. The extra capacity meant that Brain's could really look beyond Cardiff for the first time.

It meant radical changes. For years drinkers had queued up to buy their brews. There had even been rationing at Christmas. Now Brain's had to use its brains to sell beyond its traditional customers. A marketing department was set up in 1984 under Tony Smith from Express Dairies. Advertising and sponsorship were greatly increased. Lagers were introduced at the New Brewery in Roath, where a canning line was installed in 1987 to supply the take-home trade and supermarkets.

Brain's pushed back its boundaries beyond the city, and not just into wider South Wales. Through wholesalers and by targeting tenants of the major brewers (once the Monopolies and Mergers Commission in 1990 forced the big combines to allow their pubs a guest beer), Brain's ales crossed the border into England. Their reputation was helped by the success of Brain's Dark in winning Best Mild Ale at the Great British Beer Festival in London in 1991. Exports winged their way to France, Italy, Spain and the United States.

The company became involved in ambitious building projects in England, notably a £2.5 million marina development in Bath with licensed boathouse, opened in 1991. But it did not neglect its own backyard. The same year it also unveiled the grand £3 million Wharf complex overlooking Bute East Dock in Cardiff. Where Welsh coal once sailed to the world, Brain's beers were now

served in two waterside bars and a ninety-seat restaurant, with a 75ft sailing barge moored alongside. Another £3 million was splashed out on its first hotel in the capital, Churchill's.

It was spending hard to compete with the major brewers, notably Bass (which had taken over Hancock's) and Whitbread (which had swallowed Rhymney). It was a David and Goliath battle. For while Bass, through its subsidiary Welsh Brewers, accounted for 40 per cent of beer sales in South Wales, Brain's amounted to just 7 per cent, according to the *Financial Times* in 1988.

There were inevitable casualties. The lagers could not compete with the heavily advertised brands from the major breweries and were soon withdrawn, with the New Brewery in Roath, with its bottling and canning lines, closing in 1993. The red-brick buildings were demolished in 1995. The appearance of Brain's ales in English pubs tended to be temporary. Even a major TV advertising campaign in 1996 failed to establish substantial sales over the border. And it didn't always have complete belief in the selling power of its own name. When it acquired a small chain of pubs in the West Midlands in the mid-1990s, it preferred to badge them with the contrived name of Ezra Ironside's Black Country Pub Company.

But Brain's could play the giants at their own cloak and dagger game – brewery takeovers. In 1997 the company accelerated its development by a surprise swoop for its surviving South Wales regional rival, Crown Buckley. Chairman Christopher Brain later admitted the deal was the result of secret, nocturnal meetings. 'You wouldn't wish it to leak out,' he said. 'You'd be surprised at how the adrenalin keeps you going. I was driving home at four o'clock in the morning by myself and the blackbirds had just started to sing and I thought, crikey, we've just acquired Crown Buckley. That was some momentous night. That day at eleven o'clock we had our AGM, so it was all fairly tight scheduled stuff.'

He believed the two companies were 'an obvious fit' since Brain's was dominant in Cardiff while Crown Buckley's core market was in the mining valleys and West Wales. Brain's was strong in the tied pub trade, while Crown Buckley had a greater presence in the free trade, particularly clubs. The company had been created in 1988 through the merger of the Crown Clubs Brewery of Pontyclun and Buckley's Brewery of Llanelli, with brewing continuing only in Carmarthenshire.

Opposite: The £3 million Wharf pub opened in 1991 alongside Bute East Dock.

Right: Brain's New Brewery in Roath did not last as long as the Old, and was demolished in 1995.

Chairman Christopher Brain, left, and Mike Salter of Crown Buckley toast their merger with pints of Brain's Dark direct from the brewery cellars. But Mike Salter seems to be having the last laugh, judging from the server's T-shirt.

The latest merger established a new Welsh brewing giant, a company with combined sales of £60 million a year and an estate approaching 200 pubs. Buckley beers like Reverend James became Brain's brands. But not everyone was pleased. Brain's was in the firing line when brewing was concentrated in Cardiff, with the ancient Buckley's Brewery in Llanelli closed down in 1998, with the loss of thirty jobs. The Campaign for Real Ale organised a petition against the closure and then held a mock funeral at the brewery. And drinkers were soon worried by another shock development – when Brain's decided to abandon its own historic site in St Mary Street, where brewing had been carried out for centuries.

When the decision was announced in January 1999, the *South Wales Echo* canvassed local opinion. Everyone seemed to echo the view expressed by one man, 'It's a great shame. I don't see why after all these years they have to shut down the old brewery – it's a big part of the city's history.' Another called it 'the end of an era for Cardiff'. Christopher Brain admitted, 'It will be a great wrench for me to leave the premises.'

What had prompted the difficult decision was an opportunity too good to miss. Bass had decided to close its former Hancock's Brewery in Cardiff which, though little more than a quarter of a mile from the Old Brewery, was just outside the city centre. It was a larger, more efficient brewery on a much bigger 8.5-acre site. In contrast, Brain's plant was constricted and congested on just 1.5 acres and, as Christopher Brain explained, 'becoming increasingly difficult to operate as a brewery due to environmental and other conditions.' Already, in 1994, the last delivery drays had left the brewery after a distribution depot had been opened in Penarth Road. Now Brain's was also closing this depot and taking over Bass's depot in Maes-y-Coed Road.

There was an unexpected twist in the tale when a member of the Buckley brewing family, Simon Buckley, who had established a small brewery in Llandeilo in 1995, launched an audacious £68 million bid for Brain's the following month, promising to continue brewing at the Old Brewery site. He claimed to be backed by American money. But as much larger brewing groups had found in the past, the company was a tough nut to crack, with only fifty-seven shareholders, the majority of them members of the family. The offer was rejected. At the same time the brewery revealed record profits of £6.3 million.

Mike Salter, who had become managing director of Brain's after the Crown Buckley takeover, said the company could have followed the example of other firms and pulled out of brewing altogether and just run the more profitable pubs. 'But I really believe it is important to continue brewing in Wales. Brain's is about beer and brewing, and brewing in Cardiff. But the Old Brewery is a very difficult site. There's a lot of history there, which it is sad to end. But the move is a consequence of our decision to continue brewing.' Christopher Brain added, 'What we must do for our grandchildren is put the business in a position where it can be developed.'

After extensive trial brews at the Crawshay Street plant, the Old Brewery finally closed, with the last brew in September 1999. But the St Mary Street site did not fall silent, with the brewery yard being immediately converted into a huge pub for the Rugby World Cup, with more than 50,000 pints served to some 15,000 fans over six weeks. It was a fitting, final booze-up in the old brewery. Then developers moved in to create a new £25 million restaurant and residential Old Brewery Quarter, retaining some of the 1887 stone buildings. Hopes for a tiny boutique brewery at the site never materialised, but part of the yard was combined with the old brewery tap, the Albert, to create the Yard Bar and Kitchen, an extensive pub on two floors with three bars.

Brain's was going for growth in a big way. Backed by a consortium of banks with a £50 million war chest, it aimed to substantially increase its number of pubs. Many were bought in parts of Wales where it had a limited presence, such as Aberystwyth, Swansea and Monmouth. English cities like Bristol were also targeted. Modern café bar concepts like Bar Essential, first launched in Windsor Place, Cardiff, were rolled out into other areas. New catering houses, Highway Taverns, were introduced.

Brain's ensures its name is prominently displayed all over the former Hancock's Brewery.

It also pushed further out of its Glamorgan heartland by taking over beer wholesalers to the west and east; first James Williams of Narberth in Pembrokeshire in 2002 and then Stedman's of Newport in 2003. In 2005 Brain's added a further twenty-seven pubs mainly in West Wales, when it bought Innkeeper Wales of Cardigan, taking its total estate to over 250 houses. Business was booming. In 2004, the Cardiff brewer broke the £100 million turnover barrier for the first time.

Meanwhile at its new brewery, the Allbright name on the chimney stack had been replaced by Brain's. The Cardiff family brewer now ruled over Bass's last brewery in Wales, opening its new headquarters there in 2000. The much larger capacity allowed it to undertake contract brewing for the major brewing groups, while production of its own beers in 2004 topped 25 million pints (more than 80,000 barrels a year).

It now claimed to be the national brewer of Wales. It certainly was the toast of the nation. Thanks to an ambitious sponsorship deal, its name was stretched across the chests of the Welsh rugby players as they won the Grand Slam at the Millennium Stadium in Cardiff in 2005. And its beers poured down many happy throats afterwards.

But Samuel Brain's ghost must have been more satisfied that the firm had finally prevailed in Cardiff. When it moved into its new brewery, Brain's had taken over what had once been the heart of operations of its greatest rival, Hancock's.

Hancock's

The cosmopolitan nature of Cardiff is illustrated by the origins of its two leading brewers. While the Brain family breezed in from Gloucestershire, Hancock's sailed across the Bristol Channel from Somerset. Both were attracted by the opportunities in the boom town. Both established

The sign which dominated brewing in South Wales for decades.

major brewing businesses. But there the similarities end. For while the Brain's family took firm roots in the city, the Hancock family only lent their name to the enterprise and drifted in and out of the docks. Though one member did stay long enough to become a sporting legend, nearly all eventually vanished back from where they came – Wiveliscombe.

William Hancock had built a brewery in the Somerset market town in 1807, enlarging it in 1830. His son, also called William, took over on the death of his father in 1845. The substantial buildings on Golden Hill dominated the surrounding countryside. There was just one problem. In such a rural area, there was not a large enough market. William Hancock began to look elsewhere for additional trade.

Wiveliscombe was some fifteen miles inland from the port of Minehead. It was only a short sail from there to Cardiff. By 1871 the brewery was shipping its beers into West Bute Dock – into the hands of two men whose family came to dominate the company. William John Gaskell of Crockherbtown, Cardiff and his son Joseph were merchants, specialising in ale and cider. From their Collingdon Road premises they handled Findlater's Stout from Dublin and ales from the Burton Brewery of Burton as well as Hancock's. Soon they had built up 'a very considerable business' and the Golden Hill brewery was struggling to keep up with cross-channel demand.

So in 1883 William Hancock bought North and Low's Bute Dock Brewery in the docks. Joseph Gaskell and his brother John became managers of the Cardiff enterprise, which included several pubs. But William Hancock, viewing the business from Somerset, took a much wider view than Samuel Brain, brewing in the heart of Cardiff. The next year he bought the Anchor Brewery in Newport, where he had also shipped his beers since the early 1870s. John Gaskell became manager.

Trade prospered and the Welsh venture took on a life of its own, far removed from Wiveliscombe. In 1887 William Hancock & Co. Ltd was formed as a separate company, with a share capital of £100,000 and registered offices in Hancock's West Bute Dock warehouses. It was one of the earlier limited liability brewing companies in Britain, after the successful flotation of Guinness in 1886. The move meant it was a step ahead of most of its rivals in the ability to raise large-scale capital. Brain's only followed ten years later. This was vital if it was to buy more pubs – it already had forty-six in Cardiff. They were becoming increasingly expensive once the free-for-all in licences had ended during Gladstone's government of 1869-74.

Hancock's now had the muscle to sweep through the South Wales brewing industry like a sharp scythe through dry barley. Though William Hancock was chairman and local notables like Alderman Patrick Carey of Cardiff were on the board, the driving force was the Gaskell family. Joseph and John Gaskell were joint managing directors, along with William Hancock's son Frank, while Charles Gaskell was secretary.

In the days before the fine art of public relations was developed, Frank Hancock was an ideal man to generate goodwill for the new company, since he played rugby for both Cardiff and then Wales, eventually captaining the national side in 1886. That same year, he captained Cardiff Rugby Club in one of its most memorable seasons when, out of twenty-seven games, they were only defeated in their final match by Moseley at Cardiff Arms Park. They scored 131 tries, while their own line was only crossed four times. Everyone in town wanted to be linked to such back-slapping success.

His brother Philip was also a larger-than-life rugby celebrity of the period – but for the old enemy England. His first cap was against Wales in 1886. Nicknamed 'Baby', despite being an imposing 6ft 4in tall and sixteen stone, he played for Blackheath, England and the British Lions. Though always based in Somerset, he became a member of the board in Cardiff in 1902 and was briefly chairman from 1929-33.

The simplest way for the company to expand was to buy breweries. The first acquisition in 1888 was Dowson Brothers' Phoenix Brewery in Working Street, with fifteen pubs and a maltings in Cowbridge Road, for £73,000. Samuel Brain, who had just built his new brewhouse, must have felt a twinge of regret that he had missed out. This was where he had learnt his trade. Hancock's adopted the Phoenix's firebird symbol as its own and moved its headquarters from the docks to this central site.

The following year, the South Wales Brewery in Salisbury Road, with twelve pubs, was snapped up for £36,000 from Biggs and Williams. But Hancock's was also looking beyond Cardiff to the west. With the share capital now increased to £276,250, it bought Swansea wine merchants Joseph Hall in 1889 and then two breweries in Swansea – Ackland and Thomas's West End Brewery in 1890 and Thomas Jones's High Street Brewery in 1891. The Hancock name now swept the South Wales coast, with breweries in the three main ports.

But in Cardiff brewing on three different sites was an inefficient nuisance. The Gaskells must have gazed with envy at Samuel Brain's towering new brewery. It was receiving all the attention. Brewery chronicler Alfred Barnard showed no interest in sticking his head round the door of their more modest brewhouses. Hancock's was also being overshadowed by a new landmark in town – the County Brewery Company near the main railway station.

'This large and important brewing concern was founded in the year 1889, and the splendid brewery in Penarth Road is not only one of the structural ornaments of the town, but is also a very conspicuous object in the view of travellers to Cardiff, either by road, sea or rail,' boasted a commercial profile of 1893. 'The whole establishment ... is undoubtedly one of the largest and best-organised breweries in the Principality.'

Hancock's could not resist. It could smell the potential. In 1894, the company bought the County Brewery from the owner, Mr F.S. Lock, together with six pubs, and began expanding the plant so it could absorb the production from its other Cardiff breweries. New offices, stables and a bottling plant were also built. The extensive works took eighteen months. The overall cost was high – £100,000 – at a time when it was also loaning half that amount again to licensees to secure their business. But the aggressive firm could afford it, as it doubled its share capital to £545,000 by the end of 1896.

In 1895, it also absorbed another well-known Cardiff brewery, along with twelve pubs. The Castle Brewery in Great Frederick Street had been run by Henry Anthony since taking over from the Thomas family in 1862, when they moved to the Old Brewery. It had been involved in a celebrated court case, when Anthony sued a landlord for not paying for his beer. The pub denied receiving delivery. Anthony argued that the claim could be upheld by the drayman who, although unable to read or write, kept a tally in chalk marks on the cellar door. 'Send for the man and the door,' ordered the judge, telling all witnesses to remain in court. Both were brought and the case was proved.

Nothing, it seemed, could resist Hancock's financial muscle. In 1896 it even bought the wine and spirits business of Stevens and Sons in St Mary Street, where Samuel Brain had once been chairman. But the company was also unusually concerned for its workers. At its AGM that year, William Hancock announced that it had been experimenting with a profit-sharing scheme for employees.

At the end of the year, William Hancock died at the age of eighty-six. He had established a brewing empire across South Wales and, as he told the AGM in July, had concentrated all its activities within Cardiff, apart from the maltings in Canton, at one site off Penarth Road. The County Brewery had been simply renamed The Brewery, Cardiff, and was producing nearly

A Hancock's dray delivers bottled beer to an off-licence in Broadway, Roath' in 1906, including Worthington Pale Ale. At the 1896 AGM Joseph Gaskell had praised the 'excellent quality' of this Burton beer for boosting the firm's profits.

Canton Cross was the last brewery taken over by Hancock's in Cardiff. Only the pub on the site, pictured in 1927, survived.

2,000 barrels a week, ten times the production of the original plant two years before. It now claimed to be the dominant force in the borough, owning well over 100 pubs and dealing with two-thirds of the free trade. Brain's might dispute who was top dog in town, but outside Cardiff, Hancock's was the clear leader. William Hancock, ably assisted by the Gaskells, had established the largest brewers and bottlers in Wales.

On William Hancock's death his son Frank might have expected to step up into the chair. But just before his father died, the English family firm had been incorporated as William Hancock and Sons (Wiveliscombe) Ltd, and he had returned to Somerset to run the new company with his brother Philip. Instead, vice-chairman Valentine Trayes briefly took over in Cardiff, but when he died in 1900 the driving force behind the company, Joseph Gaskell, took his rightful place at the helm.

Joseph Gaskell had been an active volunteer soldier for many years. In 1910 he became Honorary Colonel of the Second Welch Brigade of the Royal Field Artillery. He was later to become chairman of the Glamorgan Territorial Force Association and was awarded the CBE for his services during the First World War. The Gaskell family's Army links were to remain strong for the whole of Hancock's history. And the Colonel planned the firm's advance across South Wales with military precision.

Having already occupied much of Cardiff, the company made only one further significant purchase in the town. In 1904, it bought the Canton Cross Brewery in Cowbridge Road with five pubs from a former Hancock's director, John Biggs. But Colonel Gaskell was looking way beyond the city walls. Hancock's pushed to the east, buying the Risca Brewery near Newport in 1902, with twenty-two pubs, and the small Hanbury Brewery in Caerleon in 1914. The company also secured three more breweries to the west in Swansea – the Glamorgan Brewery in 1901, the Singleton Brewery in 1917 and then, most significantly of all, the Swansea Old Brewery in 1927 with forty-two pubs. This was the Swansea equivalent of buying Brain's. The

Left: A 1929 Bateman cartoon captioned, 'The waiter who let a cork pop in a prohibition country', highlights the widespread evasion of the law in the USA, as Gerald Gaskell discovered.

Opposite: Hancock's eye-catching 'flagon wagon' early in 1926.

Vale of Glamorgan Brewery in Cowbridge was also seized during the First World War and the Abernant Brewery at Cwmgorse was bought in 1924 with fourteen pubs. All this expansion was funded by further increases in capital, with the share and loan capital reaching £945,000 in 1925. Most of the breweries were soon closed, with production concentrated in Cardiff, apart from the much smaller West End Brewery in Swansea.

The First World War was a difficult time for all brewers, with the threat of nationalisation and even a complete ban on alcohol in the air. Then there was the heavy loss of workers. More than 450 Hancock's employees were called away, around half the workforce, with at least twenty-eight killed and thirty-six wounded. The Colonel himself was deeply involved with the Territorial Army, while his son, Major Gerald Gaskell, served in France, Egypt and Palestine.

But the company was also active with other brewers on the home front, fighting for the trade. Colonel Gaskell told the company's AGM in Cardiff in August 1918, 'With regard to the Welsh campaign for prohibition, the brewers' organisation has held a dozen meetings up to the middle of July, attended by over 100,000 workers, and twelve strong anti-prohibitionist resolutions have been carried.' He confidently added, 'The fear of prohibition has passed away. It is accepted now by the Government that the people should have beer, that they want more beer, and in some cases stronger beer.'

His son was not so sure. Major Gaskell had returned from the war to become managing director of Hancock's in 1918. He was also a director of Scottish distillers, McDonald & Sons of Fort William. The whisky firms had been hard hit by the loss of their US export trade. Determined to face the threats to the liquor industry head on, in the early 1920s he went on an extended trip to the United States and Canada. Major Gaskell saw how prohibition was widely ignored. 'I have had lunch at a restaurant in San Francisco, where every available drink was procurable but, of course, it was not served in glasses.' He saw how law and order had broken down in Chicago. 'I stayed at the Drake Hotel and, shortly after I left, a gang of men entered the hotel and held up the guests, robbing them … The police entered and there was a regular battle, several people being killed and wounded.'

With beer sales stuck in a slump after the war, Hancock's needed to generate a little excitement of its own. To supply beer to its bottling stores in Newport, a distinctive 'flagon wagon' had taken to the roads, with its tanker shaped like a huge bottle. The company was beginning to learn how to turn heads. But it was not an American marketing wizard who was responsible, but a remarkable home-grown talent. Tommy Shell had joined Hancock's in 1887 straight from school, just five weeks after the company had been floated. He was a man ahead of his era; a master of spin before the word had been spun. A company profile in 1926 said, 'There is no-one else just like him. His perennial smile keeps him young; he will never grow up. Smart and dapper … he is everywhere … verily a live wire at Hancock's.'

Colonel Gaskell recognised his value and made him manager of the bottling department in 1902, initially on three months' trial because 'he had qualms whether Mr Shell's youthfulness was not a bar to success in such a responsible position'. Bottling at the time was the cutting edge of the brewing industry, the shining new growth area. It was just right for an innovator like Tommy Shell. He made his mark with the introduction of filtered flagon ale, bottled under a slow cold storage system. This meant the sparkling beer did not have a sediment, like most of its rivals.

But it was the way that Tommy Shell sold these new beers that really marked him out. In the highly conservative business of brewing, he took a risk. With Colonel Gaskell's support, in 1904 he placed a full-page advert in the *Western Mail*, offering to give away 3,000 four-flagon cases of ale and stout to the first 3,000 applicants. 'The following day we were overwhelmed at the brewery with postcards applying for the free gift, the number received being over 6,000,' recalled Shell. Colonel Gaskell must have choked on his whisky – a flagon contained two pints, so this was 6,000 gallons of free beer. But not wishing to disappoint eager customers, he agreed to supply all the applicants. And it paid off. 'When the agent called a few days after for the empties, he secured in almost every instance an order, and thus in one day, through a daring and generous advertisement, a large business was established,' boasted Shell.

He had made his reputation. When, in 1907, Colonel Gaskell decided to start a mineral water department under the Apex brand name, Shell was put in charge. He was even employed on a

Hancock's 1924 sailor poster by Lawson Wood was widely displayed in their Cardiff pubs.

WM HANCOCK & CO LTD TRADE EXHIBIT 1930.

This mock-up of an olde worlde tavern was used on the back of drays in street processions.

profit-sharing basis. He handled all advertising, using established artists like Lawson Wood to design posters. Colourful, eye-catching posters were, at the time, the equivalent of TV advertising today, and popular designs were widely reproduced on pub showcards and packs of playing cards.

But Shell did not stop there. He was always looking for something new to intrigue the public. He tried electric slogans bubbling across Queen Street at night. Huge neon signs were placed on the side of pubs, advertising 'Hancock's Amber Ale, 4d a bottle', with a large glowing glass. At events like the Pure Food Exhibition in the Drill Hall, Cardiff in November 1926, he introduced colossal 6ft bottles. Hancock's stand also featured the 'Ever-Flowing Flagon', in which a bottle of oatmeal stout appeared to constantly pour into a glass without it ever overflowing. The frame of a mock-Tudor 'hostelrie', Ye Hancock's Arms, was built in 1930 to be carried on dray lorries in carnival processions across South Wales. Some of Shell's measures would shock today. 'Rastus', a 4ft 9in model of a smiling black page boy, with moving head, arms, eyes, mouth and even eyebrows, was used from 1927 to hand out leaflets. When not in demand for exhibitions, he could be found in Hancock's off-licence in City Road, Cardiff, dishing out price lists.

The company also took flight from the phoenix and sought a more substantial symbol to sell Hancock's beers. A patriotic member of the Gaskell family suggested a cheery John Bull and, drawn up by the *Western Mail* cartoonist, Leslie Illingworth, it was adopted in 1926. By 1928, the top-hatted figure, raising a glass on the label, was used to promote a new bottled beer. In typical Tommy Shell style, Hancock's Nut Brown Ale was launched with a fanfare at Cardiff Races in Ely on Easter Monday.

Shell also developed one of the company's most distinctive features – its catering business. Using the Gaskell family's extensive links with the Army, from 1925 it began to supply food and drink, first for military events and local shows, and then for grander functions across the country. Its motto was 'We Cater Everywhere' and for once a Welsh brewery really broke out of

As well as catering at many shows in the 1930s, Hancock's emphasised their presence with displays by their distinctive teams of grey horses.

Wales. By the mid-1930s events included the Richmond Royal Horse Show, the Bath and West Agricultural Show, and Kent and Suffolk County Shows. Hancock's was also the official caterer to bodies like the Welsh Golfing Union and Malvern Winter Gardens, as well as local institutions like Cardiff Race Club and Cardiff Aeroplane Club. It took over rival catering companies as far apart as Dudley in the West Midlands and Brighton on the South Coast. In Cardiff, it ran the Connaught Rooms, where 750 guests could be entertained.

Hancock's also boasted that it ran 'the wine cellars of Wales'. Though it had been involved in the wines and spirits trade in Cardiff since 1896, when Stevens and Sons was taken over, it was not until the early 1930s that it developed this side of the business. The number of drinks stocked more than doubled in five years from 809 to 2,011 by 1937. Its slogan was 'You can get it at Hancock's'. The range of whiskies included its own brands Duchess and Edinbro' Cream.

A stroll through the bottling rooms with the sixty-six great black and white 'standing pieces' (huge storage casks), from which the wines and spirits were drawn off, must have been a heady experience. There was the Cocktail Room, the Liqueur Room, the Champagne Room and the Tonic Wine Room. Everything was there from cowslip and elderberry country wines to vintage brandy and champagne. There was even a special room for kosher wines for Jewish customers under the supervision of a Rabbi. The senses could also be sent reeling in the tobacco store, heavy with hints of Havana, which sold over half-a-million cigarettes alone each week.

But behind the giant 'standing pieces' lay a great gamble by Gerald Gaskell, who had become chairman in 1933 on the death of Philip Hancock. This was a high-risk strategy. For Hancock's was going for growth in the most expensive drinks at a time of the deepest depression; at a time when, from 1931 to 1936, the company was unable to pay a dividend. But the strategy paid off. Hancock's increased wines and spirits sales four times in the five years from 1932. These extra additions to Hancock's basic brewing business helped it to ride out the Depression. It was not

Lieutenant Colonel G.N. Gaskell gets an anti-aircraft gun ready to fire. The whole family was deeply involved with the military during the war.

so bent over a barrel as its rivals. Its catering arm in particular allowed it to reach out beyond the hard-hit South Wales coalfield into more prosperous England.

In 1936, as trade started to pick up again, Hancock's began to expand once more, buying up twelve Welsh pubs from Bristol brewers George's. More significantly, John Bull strode confidently up the Taff Valley from Cardiff to take over the largest and oldest brewery in Merthyr Tydfil, Giles and Harrap, with sixty-two houses. The Merthyr brewery was closed and two new floors added to the fermenting block of the Cardiff brewery to keep up with demand, with the latest aluminium vessels installed.

Another growing feature of the Crawshay Street site was its facilities for workers. The club rooms not only contained dartboards, table tennis and billiards tables, but also a radiogram, a library and an assembly room with orchestra stage for dances. But the distinctive touch was a gymnasium with punch balls and sculling machines and an air-rifle range. Hancock's teams were regular local champions and winners of the Major Gaskell Bowl. 'It is a point of honour that, although most other works teams attract outsiders, only Hancock's employees are allowed to shoot in the teams,' stressed the company. When war broke out again, this was one firm well prepared to fight.

And that meant the brewery was left struggling once more, as workers marched out of the yard to fulfil their duty to King and country. Gerald Gaskell's son Joseph, who had started work at the brewery in 1935, was as keen on recruiting for the Territorial Army as his father and grandfather. His success meant there was a manpower crisis when the Territorials were called up in 1939. His father was still just as active, commanding the Cardiff Home Guard. A key worker, brewer Ken Morison, was even called up to help run one of the more bizarre developments of the war – Davy Jones Brewery set up inside a ship, the *Menestheus*, using sea water and malt extract to produce beer for the troops in the Far East. Also, Hancock's efficient catering arm was mobilised to operate vital military and industrial canteens.

After the war the catering corps continued to advance, taking over rival companies in England, like Osmond & Sons of Salisbury in 1950. The brewing business tried to follow the knife and fork brigade over the border. It already had a few English inns, like the Lamp Tavern in Dudley, and from 1946 it leased the redundant Queen's Cross Brewery behind the pub as its West Midlands depot. Hancock's also joined the desperate drive for exports in a bid to pay for the crippling war, with boxes of bottled Export Beer being shipped out of the docks. But these advances into unfamiliar drinking territory failed to prosper and were soon abandoned. John Bull had begun to lose his swagger − and his independence.

Hancock's had long links with Worthington of Burton upon Trent, having supplied their beers in South Wales since the nineteenth century. In 1927 Worthington had merged with Bass and, after the Second World War, the English brewing giant gradually began to draw the Welsh brewing giant into its fold. Gerald Gaskell was a prominent member of the Brewers' Society, having been chairman of the society's wartime publicity committee. After the war, Bass made him chairman of its subsidiary company in London, the Wenlock Brewery. His son Joseph, who had become joint managing director of Hancock's in 1949, was also made a Wenlock director. A Bass director, W.P. Manners, was already on Hancock's board, the Burton brewers having in 1958 taken shares worth £50,000 to replace a loan.

In 1959 Gerald Gaskell died and his son became chairman of Hancock's. Bass edged a little closer, taking an equity stake in the Cardiff company of up to 25 per cent at the family's invitation and sealing a long-term trading agreement. Two years later, Joseph Gaskell was appointed to Bass's board. It was a flattering honour since Bass was one of the biggest names in brewing. He had stepped up to sup at the top table.

ON THE ROAD TO RECOVERY
-TAKE

HANCOCKS
Amber

AND KEEP GOING!

Opposite: A Hancock's dray delivers bottled beer to the docks in the late 1940s.

Right: A buoyant 1948 Hancock's advert, but the company found life more difficult after the war.

What had prompted this leap into bed with Bass was the disturbing figure of Eddie Taylor. An energetic Canadian brewer, Taylor was seeking to build a brewing empire in Britain through which to sell his Canadian lager, Carling Black Label. Through a torrent of takeovers, he had already set up Northern Breweries in the North of England and United Caledonian in Scotland, and was now prowling further south. He had an alarming North American habit of just turning up at a brewery and asking 'How much?' When he came knocking on the door in Cardiff, Joseph Gaskell jumped.

It was a cosy relationship, but the understanding was as clear as a glass of Draught Bass. If Hancock's was to be taken over, there was only one suitable suitor – Bass. And in the new land of the giants, with major British brewing groups forming in the early 1960s, the merger was only a matter of time. The only surprise was that it took until almost the end of the decade.

In 1960 Hancock's had marched out on the takeover trail for the last time, when it bought Aberystwyth brewers, David Roberts & Sons, for £680,000. The company paid the bill by issuing more shares, bringing the capital up to £1.8 million. The deal added a further 117 pubs, mainly in West Wales, to Hancock's estate of 440 houses. The Aberystwyth brewery closed, with the Crawshay Street brewery expanded again.

A major investment of £350,000 was also made in a new bottling plant. When it was officially opened by the Lord Mayor of Cardiff in 1963, it was said to be the most modern in Europe. The 18,000 sq ft hall had taken two years to build, doubling the amount of beer that could be handled. At the opening, the Mayor commented that the brewery was both a 'family affair' through the Gaskells and a 'city affair' as Joseph Gaskell and secretary Ronald Tucker were city councillors. It was a statement of solidarity with a local firm under pressure. For despite the Bass links, other groups were circling, with an envious eye on Hancock's 550 pubs in heavy-drinking South Wales. Takeover rumours had been swirling since 1962.

And the Bass connection was no longer cosy. When Joseph Gaskell had taken his seat on Bass's board, he was not sitting in the real seat of power. The Bass board was a well-upholstered illusion. For in the new world of complex combines, there was now another board above that. Bass in 1961 had merged with Mitchells and Butlers of Birmingham to form Bass, Mitchells and Butlers. To the outsider it looked like Bass had taken over another company. In reality, the more aggressive management of Mitchells and Butlers had taken control. The days of binding handshakes and gentlemen's agreements were rapidly coming to an end. What was known as the beerage – the leading families who ran the industry – had had its day.

In this harsh new world, Hancock's was running hard to keep ahead of the pursuing pack. The company was having to spend heavily to hold on to its trade. The redevelopment of town centres like Cardiff saw many pubs demolished, as back-to-back streets were cleared. And it cost considerable sums to build new houses. The company was building four to six a year, sometimes in conjunction with Bass to keep expenditure down. It was also spending heavily on the Cardiff brewery. To expand it even had to reclaim one-and-a-half acres from the River Taff, with thousands of tons of rock and 250 yards of steel sheet piling used to create a new river bank. It cost the earth to move earth on this scale. Its finances could not, however, stretch to the heavy repair costs required at its maltings in Canton, and these were closed in 1962.

In 1964 Hancock's announced record profits of £691,000, up 13 per cent. The shareholders were pleased with a dividend of over 22 per cent. But the bank overdraft had topped £1 million. To cut excessive borrowing, the company raised £900,000 through a debenture issue. But what had once been its strength – its ability to raise large-scale capital by repeatedly issuing shares – was now its weakness. With so many shares on the market, outside groups could buy control. Unlike Brain's, where the shares were controlled by the family, it had no protection. It had grown by taking over others, and was now about to fall on the same sword. It could only choose which way to fall.

Early in 1966, it saw its fate foreshadowed when Whitbread swooped to seize local rivals Rhymney. And an old adversary made an unwelcome return. Eddie Taylor, now in charge of Charrington United, was buying up Hancock's shares. Joseph Gaskell held discussions with Bass, which by now controlled a quarter of the company's capital. He didn't want to fall into the hands of Eddie Taylor – but in the end he had no choice. For in 1967, Bass, Mitchells and Butlers merged with Charrington United to form Britain's biggest brewing conglomerate, Bass Charrington, with 11,000 pubs. Next to this hungry shark, Hancock's was no more than a fish dinner. In February 1968 the board bowed to the inevitable and recommended that shareholders accept a £7.7 million offer from Bass Charrington.

Joseph Gaskell was made chairman of the new Cardiff company, called Welsh Brewers, which combined Hancock's with Charrington United's former South Wales interests, Webbs of Aberbeeg and Fernvale Brewery in the Rhondda, to provide a total estate of 750 houses. Rationalisation swiftly followed. Hancock's old West End Brewery in Swansea was closed in 1969, along with Webbs. Fernvale followed in 1970, with all production concentrated in Cardiff. Many depots were also shut. Hancock's celebrated catering arm did not fit with the new philosophy and was sold off. Only a few Hancock's beers survived, notably its popular keg Allbright. Hancock's John Bull figure was pensioned off and replaced by Bass Charrington's toby jug on a red triangle. However, the top-hatted gent kept bouncing back on advertising for HB bitter and to mark the old company's centenary in 1987.

The new trading name, Welsh Brewers, tried to give an impression of local control, but the important decisions were now made outside Cardiff, in London and Burton. Welsh Brewers did

The last pub Hancock's built in Cardiff, the Master Gunner in Gabalfa Avenue, which opened in 1967.

not even remain the name of the regional company, which became Bass Wales and West. This was a subsidiary of Bass Holdings of Burton, which was a subsidiary of Bass plc of London. Cardiff was now well down the chain of command.

But Welsh Brewers managed to present a united front and retain a strong local identity by going back to the days when Frank Hancock's rugby exploits won the fledgling company many friends. For head of communications in Cardiff, Arwyn Owen, knew just how to play the oval ball game. A former outside-half and then club secretary for Pontypridd, he had also been an area sales manager for the Fernvale Brewery when Pontypridd won the championship in 1963 under chairman Alfie John. 'He asked the brewery to put on a barrel of beer and some bread, cheese and pickles by way of a celebration,' recalled Arwyn. 'I replied, "I think we can do better than that." ' Instead he organised a major function at the Connaught Rooms in Cardiff, and so was born the annual Welsh Rugby Presentation Dinner, which came under the Welsh Brewers name in 1969 after the takeover of Hancock's. The same year, he also launched Welsh rugby's bible, the *Welsh Brewers' Rugby Annual for Wales*.

But Arwyn Owen did not stop at rugby. There was support for soccer like the Allbright Bitter Welsh Cup and sponsorship of Glamorgan County Cricket Club, plus golf and angling. Welsh Brewers even backed the popular local sport of baseball, after their man with a nose for an opportunity saw a large crowd in Roath Park watching a cup final. He spread Bass's wealth into all aspects of Welsh life, including backing brass band competitions and choral singing as the sponsors of Cor Meibion de Cymru. There was also a St David's Day annual lunch in the House of Lords.

The Cardiff brewery might have lost control of its destiny, but Arwyn Owen had made sure Welsh Brewers had firmly established itself in Wales. And as long as the Allbright name on the

chimney stack greeted visitors off the train at Cardiff Central station, it could more than compete for attention with Brain's Old Brewery, steaming away on the other side of the tracks. But Welsh Brewers had no final say over the future of the brewery. And late in 1997, Bass announced that the plant was to close in two years' time as part of group rationalisation. Bass was taking over Ind Coope's neighbouring brewery in Burton upon Trent, giving it more capacity than it needed in the UK. So it decided to close its Welsh brewery, despite having only recently re-equipped part of the Cardiff plant, which was brewing more than 825,000 barrels a year. It was grim news for the 126 workers on site. Arwyn Owen retired.

But there was an unlikely saviour waiting in the wings. Brain's decided in 1999 to buy their old rival's brewery and move to Crawshay Street from their cramped city centre site. William Hancock must have spun in his barrel, particularly since production of the one beer still carrying his family name, Hancock's HB, was moved out of Cardiff to Burton upon Trent.

But there was some rough justice in the end. For by the twenty-first century brewing had moved on from being a national business to a global industry. An even bigger whale of a shark, the Belgian international brewing combine, Interbrew, swallowed both Bass and Whitbread in 2000. This gave the Stella Artois brewer an unacceptably large slice of the British market, and it was ordered by the Government to sell off parts of Bass. These were snapped up by American brewer Coors in 2002. Among the beers the Colorado company inherited was Hancock's HB, and it moved production back to its original Cardiff home. Only now Hancock's was brewed under licence by Brain's. Samuel Brain had got his hands on his bitter rival's beer, as well as his brewery.

Opposite: Arwyn Owen, left, and head brewer Tony Duckworth sample their Centenary Ale for the Welsh Rugby Union in 1981.

Right: The brewery in 1996, a year before Bass announced it was to close.

Ely

While Brain's and Hancock's had been arm-wrestling over the bar in the centre of Cardiff, two even closer rivals had been glaring at each other for decades over the railway tracks in the outlying village of Ely, wondering who would bottle first.

The water in this rural area, which only became part of Cardiff in 1922, was ideal for brewing. Which is why this riverside hamlet, on the road west out of the city, came to house three major breweries, despite having only three pubs. The first was the Ely Brewery, the second Crosswells and the third did not brew beer at all. It was Chivers malt vinegar brewery. Its factory also produced jams, pickles and sauces. All were tightly packed into an area next to Ely Bridge, with the main railway line to West Wales running between the two ale breweries. But, as Ely always liked to boast, it was the original Ely Brewery.

The first Ely ale flowed from around 1853 after David Davies leased land from Lord Romilly to build a brewery, though there had been a maltings in the village before then. By 1863 Matthew Cross had taken over and, as sketches on his trade cards show, the buildings were substantial for a rural area. The four-storey brewhouse by the bridge would have been a major landmark on the road. By 1875 new owner James Ward was calling it the Tower Brewery, Ely. He was also proudly pronouncing in adverts that he enjoyed the patronage of prestigious customers, claiming to be brewer to both the Marquis of Bute and the Prince of Wales, and carrying their coats of arms. This probably meant he had won the honour of supplying Cardiff Castle. It was a definite three feathers in his brewer's cap. But pride often comes before a fall. By the mid-1880s the brewery had stopped brewing. The company which revived it was a remarkable venture.

Left: The village of Ely boasted another two breweries besides the Ely Brewery featured in this late 1950s advert.

Below: The Ely Brewery towered over the rural area as shown in this late 1860s trade card.

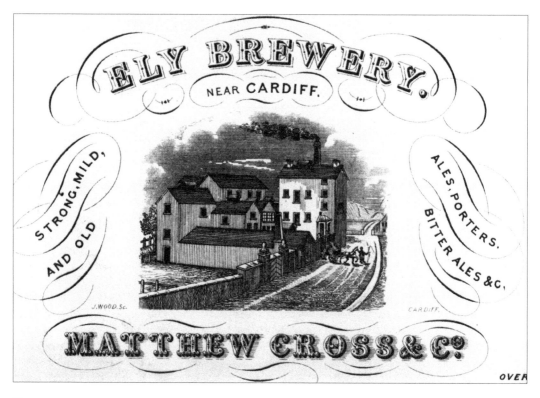

A statue was erected in 1920 in Aberdare to Caradog, twenty-three years after his death, appropriately conducting the traffic outside the Black Lion Hotel.

Unlike Brain's and Hancock's, Ely was a deeply Welsh concern. When the Ely Brewery Company Ltd was formed late in 1887, one of its directors was Griffith Rhys Jones, better known as Caradog, the celebrated Welsh violinist and conductor, who had built up the Welsh Valleys' fame in choral singing to a magnificent crescendo. In 1872 he had taken his Cor Mawr, a huge choir of 460 voices, to London to compete at Crystal Palace. The choir filled eighteen railway carriages, but it was worth the fare – Caradog's singers carried off the top prize. He returned to Wales to huge acclaim. In 1873 he repeated the feat. For a man not yet forty, who had started life as an ironworks blacksmith, it was a major achievement. He became a local hero.

The only ones who were not cheering were the committed temperance campaigners. For the musical renaissance in Wales in the second half of the nineteenth century had been built around the chapels and hymn singing. The purity of song was meant to replace the evils of intoxication. The hallelujah chorus would drive out the demon drink. But Caradog's success dampened this dry ambition. He maintained one of the happiest double acts in history. He linked beer with song. He was as happy lifting a glass as a conductor's baton. He not only went into pubs, but ventured the other side of the bar, becoming a publican in Treorchy while running the town choir. Now he was not only a musical director but a brewery director.

Caradog was not the only landlord from outside Cardiff on the board. Two other members were innkeepers from the Bridgend area – Job Llewellyn of the Blaengarw Hotel and William John Jones of the Ogmore Valley Hotel. A fourth director was William Beith of the Windsor Hotel, Penarth. The new enterprise was essentially a beer sellers' co-operative. The brewer Evan Davies was the first chairman, but died within a few months before the first beer was brewed, and was replaced by the more imposing figure of Richard Knight Prichard of Marlboro Grange, Cowbridge. Caradog became vice-chairman.

Although the company was formed just a few months after Hancock's, it had much more modest ambitions. Its £10 shares were mainly aimed at publicans. Licensees taking a stake had to agree to take the brewery's beer. There was no large-scale capital from outside the trade, no programme of major expansion. In fact, the concern was regarded with some financial suspicion. The National Bank of Wales refused to allow a £2,000 overdraft and the new company was forced to switch banks.

Costs were kept to a minimum. The venture was virtually starting from scratch, with decisions like whether to employ a salesman or buy horses and harness deferred until the first brew. The death of the brewer had been a blow, and it took six months until March 1888, to get the brewhouse in full working order. A new copper had to be ordered from Adlams of Bristol. A deeper well needed to be sunk. It was a slow start with average weekly sales in June of only sixty-nine barrels. By February 1889 it had risen to eighty-three barrels.

But the business was beginning to take shape, with two prominent pubs bought in Cardiff. The first in 1888 was the Terminus Hotel at the bottom of St Mary Street followed by the Oxford Hotel in The Hayes. With finances improving, more pubs were bought or leased outside Cardiff, as far up the Valleys as Aberdare and Maesteg. Caradog also ensured it supported eisteddfods in local pubs. By 1890 it was seeking to raise £20,000 by issuing 200 debentures worth £100 each, and was installing an extra ninety-five barrel cedar fermentation vessel to enable 270 barrels to be brewed a week, a capacity of 14,000 barrels a year.

In 1899, it made its most significant sortie into Cardiff when it bought the Ship Brewery in Millicent Street with its eleven pubs. This family firm had been run by William Phillips and then his son, also William, since at least 1848. The takeover markedly increased production at Ely from 20,166 barrels in the year ending March 1899 to 28,921 barrels the following year. It was still barely a quarter of Hancock's output, but a considerable improvement. The dividend was 20 per cent. In 1899 Ely had also adopted the Ely Bridge as its trademark, as it built bridges between Cardiff and the mining valleys. A depot was opened in Aberdare. Chairman R.K. Prichard told the 1900 AGM he was determined 'to keep up the reputation of the brewery' for the quality of its beer. It was a reputation others were keen to copy.

That year it had received an unwelcome jolt. Crosswells brewery had opened opposite, trading under the title of the New Brewery, Ely. The established concern feared customer confusion, and painted 'The Original Ely Brewery' in large letters on its Cowbridge Road wall. Battle was joined, and Crosswells daubed its own slogans across the height of its facing brewhouse. The neighbours had begun bickering. But Ely shareholders – many of them publicans – had little to complain about. From 1901 to 1911 the company's dividends remained at an annual 20 per cent. In 1912 they had even more to smile about when £35,000 from the reserve fund was distributed among them. Despite the difficulties of the First World War, the dividend only fell to 10 per cent and then revived to 15 per cent. It was a comfortable business.

But as it pushed for trade up the Valleys, it bumped into its much larger, more aggressive big brother. Like Ely, the original driving force behind Rhondda Valley Breweries had been licensees, when it was established as Rhondda Valley Brewery in Treherbert in 1873. Nearly twenty years later, in 1892, it bought the oldest brewery in Pontypridd, the Pontypridd Brewery, and in 1896 Rhondda Valley Breweries was set up to combine the two businesses with ninety-seven pubs. This was a much bigger company than Ely, and it kept growing, taking over Pontypridd United Breweries in 1918. This made it five times the size of Ely, with assets worth over £1 million in 1919, compared to Ely's £219,431. That year its profits were £43,250 compared to Ely's £6,624. Ely might be the brewery from the big city, but it was no match for this giant of the coalfields, and in 1920 the boys from the black stuff moved into Cardiff in a big way.

Ely Brewery in 1906, proudly claiming along its wall 'The Original Ely Brewery', with Crosswells Brewery to the left.

Rhondda Valley Breweries was five times the size of Ely.

THE

PRIZE

RHONDDA VALLEY & ELY BREWERIES LTP

The combined group of Rhondda Valley and Ely Breweries was recording better profits than Hancock's – until it was hit by the General Strike in 1926.

Rhondda Valley Breweries did not only take over Ely, the same year it also bought control of the Cardiff Malting Company in which Brain's had once had a substantial stake. Power had shifted to the Valleys, with Pontypridd's famous arched bridge replacing Ely Bridge as the firm's logo. The combined company, which now went under the tortuous title of the Rhondda Valley and Ely Breweries, was described as 'one of the largest and most progressive companies in Wales and the West of England'. It was even more successful than Hancock's, with profits in 1924 of £131,984, double the £65,876 rolled out by its rival.

This booming business was built firmly on coal, with four out of five of its 284 pubs up the Valleys. Miners were famous for their thirst. But this reliance on one industry was also a weakness. When the General Strike in 1926 closed the coalfields for eight months, profits plunged faster than a bottle tossed down a pit shaft. As the bitter dispute was followed by the Depression of the 1930s and mass unemployment, the company never recovered. No dividend was declared for an astonishing twenty years from 1927 to 1946. The shares became regarded as almost worthless. The company also lost its leader when chairman Major John Griffith Jones, the son of Caradog, died in 1928. A famous soldier and huntsman, he had been a member of the governing body of the Church in Wales. More crucially for the brewery, he had also been an accountant by profession. Such skills were needed, as disgruntled shareholders, used to years of fat dividends, alleged mismanagement. Thomas Skurray, a leading figure within the Brewers' Society and a director of the neighbouring Hereford and Tredegar Brewery, was brought in to sort out the mess.

Drastic surgery was applied to the sprawling brewing empire, which included large breweries in Treherbert and Pontypridd as well as Ely. The Cardiff arm faced the axe since it was not at the heart of the company's operations. The more central Valleys town of Pontypridd was also home to the important bottling plant. But Ely escaped, thanks to the fact its plant had just been enlarged and modernised in 1925, plus the strength of its reputation – 'Ely Ales, Best in Wales', claimed its slogan.

Instead, the Treherbert brewery was closed in 1928 and the one in Pontypridd a year later, with all production and the bottling plant moved to Ely. The company's name was also cut back to just Ely Brewery. But the brewery was still in deep trouble. Ely's debts to Barclays Bank in 1936 added up to over £100,000. It was not so much on the floor as almost underground. There was desperate talk of a merger with its near neighbour Crosswells, but it came to nothing. Only the eminence of Thomas Skurray within the industry kept the company alive. He was chairman of the Brewers' Society from 1929 to 1931. He was also the chairman of two major English breweries and a director of many more. He was involved in politics as well as business, and lived in Berkshire where he had been chairman of the county council. He had little time to devote to Ely's pressing problems. The man who eventually replaced him, after Skurray's death in 1938, was to have much more attention to detail. His name was Lazarus Nidditch, and he was to raise Ely from the dead.

Few men can have had such a galvanising effect on a business. He was appointed financial director late in 1946 and two months later chairman. In 1949 he also became managing director. He was the right man at the right time. For towards the end of the war speculators had begun to take a fresh look at Ely. They spotted light at the bottom of the pit shaft. The war had brought an odd prosperity to the South Wales coalfield, with increased demand for steel and coal. In turn that meant more demand for beer. The *Sunday Express* described the firm as potentially 'a £2 million company'.

Nidditch soon shocked shareholders in 1947 – with a dividend. It was only 2.5 per cent, but the first for twenty years. The next year the dividend was 10 per cent with profits a healthy £238,613. The shares shot up in value. But Nidditch was much more than a miracle worker in the accounts office, as the *Joint Stock Companies Journal* recognised in 1953:

Without doubt much of the credit for a complete metamorphosis of both balance sheet and profit and loss account has to be given to this restless executive who gets through a tremendous amount of work, finding time to visit the inns and hostelries owned by the brewery throughout the Rhondda and South Wales, exercise his influence on customer relations generally, and initiate and constantly develop a spirit of harmony and cheerful mutual effort amongst the staff.

He lifted morale across the company, starting with a welcome bonus of three weeks' wages to all staff in 1948. A '25' club was formed the following year for long-serving employees, the first of its kind in the area. Another novelty was suggestion boxes, with cash awards for good ideas. A welfare hall and canteen were built at the brewery, sports clubs encouraged and social outings arranged. A house magazine, *Mild & Bitter*, was introduced in 1952.

There was also considerable investment in the brewery. Much of the equipment was replaced, many vessels being worn out after more than thirty years' service. A new floor was added to the bottling stores with the latest handling machinery. Nidditch boasted to the 1953 AGM, 'Ely Brewery is in the forefront of breweries in the use of modern methods with modern plant of copper and stainless steel, instead of leaky, wooden fermenting vats thirty-forty years of age, a worn boiler fifty-five years old and other decrepit plant breaking down weekly.'

He also worked hard to improve Ely's estate. He visited 150 of the company's 250 blue and white houses in his first year and was appalled by their 'deplorable condition of disrepair through terrible neglect'. Helped by his wife Matilda, who came on the board from 1951, he tried to replace the crude 'vertical drinking' bars with more homely lounges with comfortable seating where men could take their wives. Electric lanterns were fixed outside to brighten the post-war gloom. His marketing was just as innovative. He dismissed the old logo of a bridge as having 'no relation to our products' and replaced it with a large Ely barrel. In 1948 a film was made about the brewery for showing in local cinemas. A cartoon character called Tom Tapster, who looked

Satisfaction in every glass

ELY ALE OF COURSE

Left: Ely Brewery recorded its first dividend for twenty years in 1947; no wonder the face in this 1948 advert can't stop grinning.

Below: Ely invested heavily in its run-down brewhouse. Here foreman Harold Greatrex checks the depth of the grain in the mash tun in 1956.

Opposite: Lazarus Nidditch raises a glass at the opening of the New Ely Hotel in Cardiff in 1956. His wife Matilda, left, enjoys the moment.

like a barrel, was used to promote the company's beers. A range of bold bottled brands were introduced, led by Brewer's Own.

Ely's revival and increased standing within Cardiff is illustrated by two events in the mid-1950s. In October 1955, Ely was asked to supply the bars at the Jubilee Ball at the City Hall to mark the fiftieth anniversary of Cardiff becoming a city and the imminent announcement that it was to be the nation's capital. It was a special honour and Ely greeted the 800 guests with proud displays of its beers. A few months later, in March 1956, Lazarus Nidditch officially opened the New Ely Hotel in Cardiff, the first completely new house to be built by the brewery for decades. It was erected on the site of the Colbourne Hotel in Coburn Street, Cathays, which had been destroyed in the wartime blitz with the loss of fifteen lives in neighbouring houses. A plaque in the new lounge concluded with the words 'Victory over Tyranny'. But this was no standard, modern pub. As well as a lounge and large public bar with skittle alley, it had a room for receptions, a gentlemen's bar and – thanks to the influence of Mrs Nidditch – a unique 'ladies only' bar. There were also three flats to let.

Lazarus Nidditch relished fresh ideas. He liked sharp quotations. On the rear of the annual report in 1948 was printed, 'There are no hopeless situations – there are only men who have grown hopeless about them.' The messages had a hard edge. His autocratic approach meant someone showing initiative could be promoted on the spot, recalled a colleague. But an individual crossing him in the morning could find the chairman himself unscrewing his nameplate off his office door in the afternoon. He had no time for those who got in his way.

And while this individual style worked wonders in the brewery, it could upset vested interests outside and earn him powerful enemies. As early as 1949 he was in trouble for slating Pontypridd

licensing justices. But when he began to attack senior figures in the London Stock Exchange, in highly personal tirades, claiming they were undervaluing Ely's shares, he provoked retaliation. His main target, the influential Esmond Durlacher, refused to handle Ely's shares and the rest of the dealers followed suit in 1950. Frozen out of the City, this damaged the brewery's ability to raise capital and caused some shareholders to challenge Nidditch's rule. But his successful profits record meant he was almost impossible to topple.

But the issue became a long-running sore, with the dealers demanding an apology and Nidditch refusing to say sorry. Instead, by 1953, he had raised his gun sights and was blazing away at the chairman of the Stock Exchange, John Braithwaite, for not taking steps to get Ely share deals going again. The City editor of the *Daily Express*, Frederick Ellis, expressed some sympathy, though he described both sides in the dispute as 'childish'. It was a mistake. Immediately Ellis's desk was hit by a thumping 578-word telegram. 'Mr Nidditch's telegram is an angry, violent, threatening and accusing document. Such is the fiery spirit with which he fights his case,' Ellis told his readers. 'I spare you the details'. More lengthy missives 'of little wisdom' followed, until by the end of the year, Ellis had awarded Nidditch a mock honour for services to the GPO for 'helping reduce the losses on the telegram service'.

The eccentric episode illustrates Nidditch's volcanic character. Another event that year shows a different side to his personality. Staff presented him with an onyx desk set prior to the 1953 AGM for his 'never-failing kindness, consideration, good fellowship and sympathy'. He was a magnetic man who attracted both devotion and dislike. He was certainly impossible to ignore.

When Cardiff LVA held its annual banquet in 1954, the brewery provided guests with a plan for a new drinking mug. Its many features included a short-measure alarm, a sobriety indicator, a horn to sound when empty, a six-pip warning at five minutes before closing and a miniature gramophone underneath for playing drinking songs. Shareholders were less amused by odder adventures, like Ely taking over the World Natural Sponge Suppliers (of which Nidditch was chairman). There were also rumblings about 'Nidditch nepotism' since he and his wife had been joined on the board by their son-in-law, Dr Saper, and brother-in-law. And the discontented had found a formidable voice in Cardiff accountant Julian Hodge, who was to prove Nidditch's nemesis, by hounding him into an unwise deal.

Julian Hodge had sprung to prominence early in 1956 when the Rootes Group was taking over Singer Motors. He felt the offer was not enough. As the *Sunday Express* put it, 'He bellowed lustily for more – and got it.' Having made a name for himself as a City troubleshooter, through his company Investors Protection Facilities Ltd, he now turned his attention to Lazarus Nidditch. In a circular headed 'Goodbye Mr Nidditch', he called on the chairman, his wife and another director to step down, and forced an EGM in July 1956 to try and establish a new board. It promised to be a tempestuous encounter.

'At times there was pandemonium, a number of men being on their feet and shouting in competition,' reported the *Western Mail*. A water jug was sent sprawling across the directors' table. There was almost serious injury. 'I went up to the table to use the microphone and I grabbed it to emphasise a point. Then the loudspeaker fell from the ceiling, just missing Alderman Ferrier,' Julian Hodge told the *South Wales Echo*. Lazarus Nidditch banged his gavel and shouted, 'It's a conspiracy, a plot.'

The stormy meeting ended with the chairman ruling the motion to remove the directors invalid, but a card vote was held anyway. When the poll result was announced two days later, this backed the board by around 2.5 million to 2 million against. Nidditch crowed that Hodge had been 'well and truly licked'. But though he had won the battle, the war was far from over. His plans for Ely to take over his sponge company had been defeated, and he was clearly rattled by what he called 'the vile and bestial attacks made upon him'.

Ely was no longer a quiet, rural village by the river in the 1950s, as this swan was about to discover while admiring the brewery. Nidditch was also running into problems with awkward shareholders.

Hodge kept sniping away through circulars to shareholders, wearing his older opponent down. Though the brewery was performing well – Nidditch claimed in 1957 that Ely showed the greatest increase in profits in relation to the number of pubs and capital employed of any British brewery – the attacks on his handling of the company hurt. 'During 1957 there has been a continuation of the vindictive hounding campaign to stir up ill-will and prejudice against the chairman and to bring his name and that of members of the board into disrepute,' complained the directors' report in 1958.

By 1959, Nidditch was tiring of the struggle. He told the AGM in January, 'I say again that if anyone wishes to take over your company, considering the achievement resulting from virile leadership and management effort and success, then they should do so in an honourable way by making an open bid at a fair price, or leave us and our shares alone.' He was looking for a way out, and his eagerness to depart meant he failed to check out the men behind the money, when what looked like an attractive offer arrived. It was to prove a costly mistake.

A London finance corporation, H. Jasper and Co. Ltd, made a £1.75 million offer for Ely Brewery in August 1959. The takeover also embraced the Cardiff Malting Company, in which Ely had a substantial interest. The Jasper group had just bought a hotel chain, R.E. Jones Ltd, which included three prominent hotels in Cardiff, the Angel, Sandringham and Philharmonic. Nidditch immediately recommended acceptance. His attitude was hardly surprising since Jasper was also offering to pay him £30,000 for loss of office. His son-in-law Dr Saper was to receive £10,000. Jasper would even buy Nidditch's house, Two Pines at Creigau, for £17,000. It was an unusual arrangement, which the *Financial Times* queried. Even stranger was the Jasper group.

Harry Jasper had escaped from the Nazi persecution of the Jews in Berlin in the 1930s. His finance and property group was a ramshackle affair. In three years from 1956 he bought a string of unrelated businesses including Cotton Plantations, Capital News Theatres, Blindells shoe shops, Rubens-Rembrandt Hotels, Lintang Investments, Victory Real Estate and even Temperance Billiard Halls. In each case the main aim was to strip down the business and rapidly sell or develop the properties. It was a property empire built on unreliable foundations.

On 14 September, the deal seemed to be sliding smoothly to completion. Lazarus Nidditch and the Ely board resigned at a special meeting and were replaced by Harry Jasper as chairman and

Friedrich Grunwald as managing director. Then reality set in. There were not enough funds to pay for the bid. The value of Jasper shares collapsed, £6.4 million being wiped off in three days. Mr Jasper claimed the bid had been made on behalf of Mr Grunwald, who had disappeared without providing the money. He later turned up 'on holiday' in Israel, claiming to have had a nervous breakdown.

The Stock Exchange ordered that dealings in all fifteen Jasper companies be suspended and the Board of Trade appointed an inspector to investigate their affairs. Nidditch tried to regain control of the brewery. 'I'm the best man to put things right,' he told the *Empire News*. But it was too late. His reputation had been badly damaged. City of London Fraud Squad detectives even came to South Wales to interview him and other directors on the day before the crucial shareholders' meeting. The lengthy questioning continued until three o'clock in the morning. The same day Leslie Lowe, the chairman of a Cardiff engineering company, Davies, Middleton and Davies, which had built the Empire Swimming Pool, was appointed temporary chairman of the brewery.

The packed shareholders' meeting on 30 September at the Park Hotel in Cardiff was seething with suspicion and hostility. Nidditch could no longer rely on solid support from publican shareholders, with many bitterly questioning his pay-off of £30,000. 'If you sold your company for thirty pieces of silver, what are you doing here?' demanded one. The front page splash headline in the *Western Mail* that morning had not helped his cause: 'Fraud Squad in Wales – Detectives call on former Ely chiefs.' The paper's diary column, 'Wales Day by Day', best captured the vicious vendetta in the Park Hotel:

From the moment of Mr Lazarus Nidditch's ceremonial entrance, to the craning of necks and the flash of camera bulbs, it was one of those vintage scenes, with raised hackles and stormy words, that one has come to associate with meetings of the Ely Brewery shareholders. There was Mr Nidditch, a short, proud man with bald, nut-brown head, sitting behind the green-baize table, his rimless spectacles glinting under the chandeliers. On his left, the impassive, expensively-dressed figure of Mrs Nidditch, whose feminine touch once helped re-glamorise the Ely Company pubs.

Mr Nidditch would not claim that public speaking is his strong point. His words came in bursts of excitement and rhetoric. In spite of his inscrutable front, he is not a man who can conceal his emotions. [He had good reason to be upset. Waiting to twist the knife was his long-time opponent Julian Hodge.] It was soon evident that the main struggle of the meeting was not so much between Mr Nidditch and the shareholders, but between Mr Nidditch and Mr Julian Hodge, the financier, who has been issuing advertisements in the name of Investors' Protection Facilities about this affair.

Yesterday phrases like 'Fraud Squad', 'double-cross', 'fait accompli' and 'completely untrue' fell like angry grape-shot from Mr Nidditch's lips. While before him, as though in the centre of the arena, sat a tight, confident group, headed by Mr Hodge, all well-armed with detailed documents. At every awkward question from the floor, Mr Hodge – a handsome, upright, greying man, with all the poise of a seasoned warrior – would bare his gleaming teeth in a triumphant smile. And whenever he or one of his companions posed a question, Mr Nidditch would crouch forward and snap out a reply in a tone of undisguised vehemence.

Trapped and taunted, he eventually left the meeting early, after attempts to re-elect him as chairman failed. He had hoped to take the reins of power again, instead a more accomplished horseman jumped in. Colonel Harry Llewellyn had made his name as an Olympic gold-medal-winning showjumper at Helsinki in 1952 on Foxhunter. But he was also a leading Welsh industrialist and chairman of Rhymney Breweries, which already owned Ely's close rival Crosswells.

Lazarus Nidditch leaves the turbulent shareholders' meeting at the Park Hotel in Cardiff a beaten man.

On 21 October, Rhymney bought 900,000 Ely shares for £200,000 from Harry Jasper. These had only recently been sold to Jasper by Nidditch. Just over a week later, it followed up this 12.5 per cent stake with a full offer for the company of £1.75 million in Rhymney shares. 'It is an attractive offer and, if accepted, means that Ely stockholders will still have a holding in a Welsh brewery,' said Colonel Llewellyn. Ely's battered and bewildered shareholders were delighted.

Lazarus Nidditch refused to go quietly. He issued a bitter eight-page circular on 17 November justifying his position, which he personally hand-delivered to the *Western Mail*. 'The Jasper failure was no fault of mine and was a shock to the whole country including many astute financiers,' he argued, while attacking the 'offensive, distorted and untrue' comments made about him. He questioned the legality of Mr Lowe's position as chairman and threatened legal action. He did, however, give his 'blessing to the proposed marriage' with Rhymney Breweries. This went through in December 1959, bringing together two large Welsh breweries to create what was claimed to be the largest brewing company in South Wales, with a total of 730 pubs.

But the merger spelt the end of Ely. Its barrel trademark was soon broken up and trampled by the Rhymney hobby horse symbol. Most of the Ely beers disappeared. Within three years the Ely Brewery itself was demolished, with all production in the city concentrated at an enlarged Crosswells plant, which was officially opened early in 1963. Ely's 100ft chimney was demolished brick by brick by a man who had helped to build it in 1939. Crosswells, the Cardiff arm of Rhymney Breweries, had finally triumphed over its near neighbour.

Once Rhymney had taken over Ely Brewery, top, in 1959, it owned two breweries within a few hundred yards of each other, as Crosswells, bottom, was just the other side of the railway line. One was bound to shut.

Crosswells

Ely's neighbour knocked late on the brewhouse door. By the time Crosswells staggered into the brewing business in Cardiff, most of its rivals were well established. And it survived as an independent company for less than four decades. The company's stumbling start did not bode well for the future.

Like Brain's and Hancock's, Crosswells was an English invader. The original Crosswells Brewery had been established in Oldbury, near Birmingham, in the 1870s by Walter Showell. In 1887, the company was registered as Walter Showell & Sons Ltd. The Showells had ambitions to expand trade across a wide area. They opened depots in Bristol, Newport and then Cardiff, at No. 11 Penarth Road, in 1892. The stores, trading as Crosswells Ltd, bottled not only Showell's beers, but also Bass and Guinness, as well as supplying wines and spirits. Business boomed.

Sales were so good that pressure grew to brew on the spot, rather than trunking beer down from the West Midlands to South Wales. So in 1897 Crosswells Cardiff Brewery Ltd was formed. Despite the name, the initial intention was to brew in Caerphilly. The new company combined Crosswells Ltd of Cardiff and Bristol with the Caerphilly and Castle Brewery, Caerphilly, with maltings and eighteen pubs. Also involved were wine merchants Carey & Co. of Queen Street, Cardiff, and four hotels – the Lord Wimborne, Ruperra and Avondale in Cardiff and the Wingfield, Llanbradach.

The share capital was £140,000 (soon rising to £200,000) and the directors, under the chairmanship of Walter Showell junior, included Alderman Patrick Carey, a leading figure in the licensed trade in Cardiff, who had briefly been on the board which founded William Hancock &

Above left: Showell's of Birmingham was expanding across the UK, with pubs in London and another brewery in Stockport, as well as a bottling store in Cardiff.

Above right: The Ruperra Hotel in Splott, Cardiff, was one of the pubs involved in forming Crosswells Brewery.

Co. Ltd ten years before. It was a substantial combination, with the properties and plant involved valued at over £250,000. But from the beginning it ran into trouble.

First there were costly delays. When the third AGM was held in Cardiff three years later, the company had still not started brewing. Managing director Frederick Richards, in the absence of the ill chairman, told shareholders that the original intention to extend the brewery at Caerphilly had not gone ahead 'for several reasons'. Instead they had sought 'a more suitable site', and 'after lengthy negotiations the directors arranged to take a plot of land near Ely Station, the water supply having been proved.'

Leading brewery architects, Arthur Kinder & Son of London, were commissioned to build 'one of the most up-to-date and perfectly equipped breweries in the country' with its own railway siding into the yard. But, as the *Brewers' Journal* revealed, 'all the buildings have been kept as plain as possible to avoid any unnecessary expense on external design'. The plant inside the six-storey brewhouse was also designed for economy.

The company was already counting the pennies. For it was costing 4s 6d a barrel in railway carriage to bring 4,000 barrels a year down from Birmingham. That added up to a tidy sum. And the delays continued. In May 1899, the *Brewers' Journal* said the work should be completed in ten to twelve months. Eighteen months later, at the third AGM, it was expected to be up and running 'in the course of a few weeks'. In the end the New Brewery, Ely, did not start brewing until early in 1901.

But if the company thought its teething troubles were over, they were only just beginning. Profits were progressing, rising to £23,075 in 1902, compared to £13,060 in its first year, with a dividend of 5 per cent. But behind the scenes were major problems. In January 1904 'a sensation was created' when chairman Walter Showell's brother Charles, who had been chairman of the original Crosswells Brewery at Oldbury, was arrested on charges of conspiracy to defraud the company and falsifying accounts.

A shareholders' investigation had discovered 'bad management in nearly every department' since 1899 and a deficiency of almost half a million pounds by 1904. Manager Frederick Richards, who was also MD in Cardiff, was held on the same charges. The investigating committee concluded, 'The downfall of the business was due largely to gross mismanagement and reckless extravagance.' Properties were overvalued and profits 'grossly overstated'. Both men pleaded guilty and were jailed for fifteen and nine months each. The industry was shaken. Charles Showell had been an eminent figure, a former chairman of the Country Brewers' Association. Shares in which he had been involved collapsed. Crosswells in Cardiff was caught up in the shock waves and took years to recover. No dividend was paid for seventeen years from 1904 to 1920.

Only in 1927 did Crosswells feel confident enough to leave its troubled past behind and make a significant acquisition, when it bought the central Cardiff business of William Nell in St John's Square, with twelve pubs. The family company had run the Eagle Brewery next to the Tennis Court Hotel since 1846. William Nell was also a prominent member of Cardiff Town Council. On his death in 1871, his son William Walter Nell took over. In 1890 he turned Nell's into a limited liability company valued at £65,000, but the true worth of the concern was in its extensive cellars not its brewhouse. Nell's was best known for its dealings in wines and spirits and also handled many other beers, including London and Irish stouts and Burton and Scotch ales. It was a noted importer and ship supplier, an important trade in the port of Cardiff. Brewing always seems to have been a sideline, although it did own a maltings in Bridgend and in 1898 had taken over the small Abergwawr Brewery in Aberaman, near Aberdare.

There was also a strong voice from Ely on the board. Major John Griffith Jones, the chairman of Ely Brewery and son of Caradog, had been chairman for many years, with Ernest Nell managing the business. In 1926 Nell's made a profit of just £1,361. No dividend was paid and two directors died. Many expected Ely to take over, but the coalfields giant was struggling with the effects of the General Strike and Major Jones was nearing the end of his life. He died in 1928. Instead, Crosswells swooped to thwart its rival in August 1927.

Unlike Ely, Crosswells had managed to remain profitable in the 1920s, paying a dividend throughout the decade despite difficult trading conditions, even if it did slip from 15 per cent in 1925 to 5 per cent in 1929. This was partly due to its greater trade in the more prosperous city and the broader base of its business, with more dealings in wines and spirits. The acquisition of Nell's strengthened this position.

Crosswells' chairman in the 1930s was one of the great characters of Cardiff, a shrewd if not ruthless businessman; a man who loved to gamble and make money. He might have had a grand title, Baron Glanely of St Fagan's, but this was no well-mannered aristocrat. This was Bill Tatem, one of the roughest, toughest men to come out of Cardiff docks. Born in Appledore, North Devon in 1868, he was briefly a sailor before a shipwreck threw him ashore at Cardiff, where he became a junior shipping clerk. By 1897 he had bought his first boat. Soon this had expanded into a fleet and his own shipping company, Tatem Steam Navigation. A master in the dark and dusty arts of coal cargoes, he made a fortune and many enemies. But no one could stop this wild tiger of the bay. He rose to be chairman of the Cardiff Shipowners' Association and a director of many businesses. In 1911 he was made High Sheriff of Glamorgan and in 1918 became Lord Glanely for his contribution to the war effort, although some allege it was because he presented a cheque for £50,000 to Lloyd George's political fund.

WILLIAMS NELL'S

BEER & PORTER BREWERY,

Bass', Allsopps' and other India Pale and Burton Ales, in Cask & Bottle.

London and Irish Stouts, Scotch Ales, Cider, Perry, Schweppe's Lemonade & Soda Water.

CARDIFF.

SHIPPING SUPPLIED.

Nell's was not just a brewer, but also supplied ships with a wide range of beer, wines and spirits, as this 1858 advert shows.

Tatem enjoyed the trappings of power and particularly the racecourse, where he was just as calculating as in business. He bought a stables in Newmarket and in 1919 entered two horses in the Derby. He praised the merits of one, Dominion, while quietly backing the other, Grand Parade, which won at 33-1. He made a packet while all the dock workers in the Packet pub, who had taken his tip, tore up their betting slips in fury. 'Old Guts and Gaiters', as he was known behind his substantial back, went on to win six classic horse races over the years and pocket more than a quarter of a million pounds in prize money. In 1928 he was granted the rare honour of the freedom of the city, and two years later his bust was placed in Cardiff City Hall.

This was the bold buccaneer who now headed Crosswells Brewery. If he could spot a sharp business deal, he would take it. Crosswells was an attractive target for larger but struggling breweries like Ely, whose sales depended on the depressed demand for beer in the mining valleys. In January 1935, the *Western Mail* reported Stock Exchange rumours of a possible merger. But if talks took place they came to nothing. Instead another coalfields giant made its first major advance into Cardiff.

Like Ely, Andrew Buchan's Rhymney Brewery, which had been established by the Rhymney Iron Company in 1839, had been struggling. But by 1936 sales were starting to improve, and it was in this more confident climate that it made a move to take over Crosswells. Some commentators claimed that 'recovery can be hastened by the formation of more powerful units'. Rhymney believed this. Earlier in 1936 it bought the last major brewery left in Merthyr Tydfil, David Williams' Taff Vale Brewery with some thirty pubs. Then at Crosswells' thirty-ninth AGM in October 1936, at its Cardiff offices in St Mary Street, Lord Glanely announced that he and the other directors would be stepping down 'in consequence of the control of the company having

Above left: Rhymney's distinctive hobby horse logo rode in to take over Crosswells in 1936.

Above right: Crosswell's own beehive logo soon vanished after the takeover.

passed into other hands'. Rhymney chairman Lieutenant Colonel G.L. Hoare then took over. The men from the Valleys had moved in, and Glanely left smiling all the way to the bank.

But though Crosswells had lost its independence and was gradually to lose its identity, as the Rhymney hobby horse symbol galloped into prominence, the brewery was to prove a surprising survivor. It was to last longer under Rhymney, and later Whitbread, than it had done as an independent company, brewing for a further forty-six years. The brewery built on a shoestring in 1900 even survived the merger with its near neighbour Ely in 1959.

This was due to two major factors. Having a brewery in Cardiff gave Rhymney a much higher profile and added prestige. It proved they had really arrived. Investment tended to be poured into the Cardiff plant. Notably, it became the group's showpiece bottling centre. In contrast Rhymney's own brewery was difficult to develop. When Whitbread in the 1950s produced a report on the plant, it condemned Rhymney as 'an old-fashioned brewery on the side of the hill which had a shortage of water in a dry summer. The fabric is old and leaking ... rather Heath Robinson and definitely cramped.' Crosswells in comparison was proudly described by Rhymney in its 100th anniversary booklet in 1939 as 'a modern brewery'.

By its centenary, Rhymney owned 362 pubs and it continued to expand after the war, taking over the Welsh pubs of the Hereford and Tredegar Brewery and then the Llanfoist Brewery near Abergavenny in 1945. But as it grew in Wales, it was also drawn into the larger brewing empire of London brewers Whitbread. It proved a dangerous liaison. Eventually the hobby horse was swallowed by the more powerful hind's head.

Above left: Whitbread's bottled beer stores in East Canal Wharf in 1927 next to the railway line, with Sentinel steam wagons ready to roll.

Above right: Whitbread relied on the railways to deliver its beer across the country.

Whitbread had a long history in Cardiff. Unlike many other English brewers, it kept control of its Welsh operation rather than acting through agents. It had been one of the pioneers of bottled beer and in the 1890s began to push sales outside London, opening bottling depots across England. In 1894 it opened its first in South Wales in Cardiff. The beer was brewed in the company's Chiswell Street Brewery in London and then delivered by rail in huge casks for bottling locally. Its Cardiff depot was right next to the Great Western Railway line in East Canal Wharf, with the railway bridge over the canal used to promote its beers. Despite a narrow arched entrance next to the Duke of York pub, the premises were quite extensive.

The London company also built up a chain of Whitbread off-licences in Cardiff. Many were extensions to general stores, such as Maria Taylor's shop (later Turner's) in Constellation Street, Adamsdown, which had an off-licence added in the 1890s, or Griffin's Stores in Wyndham Crescent, Canton. One of the best known, on the corner of Union and Little Union Street, was originally a pub, the Earl of Windsor. It was demolished in the late 1970s to make way for the St David's Centre.

Eventually the bottling stores' closeness to the railway proved a problem. Great Western compulsory purchased the site in order to widen the line into the main station. A new depot was built in Penarth Road and operations transferred there in 1932. This was the first of a new generation of bottling stores opened by Whitbread and a showplace for the latest machinery. Within four years the buildings had to be extended. After the war the depot pioneered one-trip bottles (bottles to be used only once) for ships, including huge orders for the liners *Queen Mary*

Whitbread's later depot in Penarth Road, just a few months before it closed in 1967.

and *Queen Elizabeth*. In the 1950s it became a centre for bottling Mackeson milk stout. In 1963 the bottling plant was closed, as bottling was switched to Crosswells Brewery, and four years later the depot was shut, after Whitbread finally took over Rhymney Breweries.

Rhymney's close association with Whitbread began in 1951, when the London brewer's energetic chairman, Colonel Bill Whitbread, joined the board at Rhymney's invitation, along with another Whitbread director. Later another Whitbread man, Frank Jupp, became Rhymney's vice-chairman and managing director. There were two main reasons why Rhymney welcomed Whitbread. After the war property prices rose rapidly and speculators began to prowl around breweries, attracted by the large number of pubs they owned, many on prominent sites. Rhymney felt vulnerable, particularly since it was losing its chairman of twenty-three years, Lieutenant Colonel G.L. Hoare, who was leaving in 1952 to become a partner in his family's bank. Colonel Bill was a reassuringly formidable figure to have at your side. The company felt it made sense to let Whitbread take a share stake and to sell its popular bottled beers alongside its own, in return for protection. Rhymney was the first brewery to seek shelter under what later became known as the Whitbread umbrella.

There was a second reason why the new chairman, Colonel J.D. Griffiths, was delighted to have them on board. Whitbread were regarded as one of the leading technical brewers of the day. Whitbread's second brewer at Chiswell Street, John Wilmot, became a director and trouble-shooting brewer, first at Rhymney and then Crosswells. This co-operation in Cardiff culminated in 1958 in the opening of a new bottling plant by the High Sheriff of Glamorgan, capable of filling 400 bottles a minute and more than doubling the amount that could be produced. Not everything went smoothly. The long lines of bottles had rattled into action before the speeches began, forcing the High Sheriff, Squadron Leader H.G. Lewis, to shout above the din. The breathless dignitary was grateful for a beer afterwards.

Significantly, among those present at the noisy opening in June was Olympic showjumper, Colonel Harry Llewellyn, who had just been appointed a director. A month later he was to become chairman of Rhymney Breweries at the invitation of Colonel Whitbread. A new twenty-five-year trading agreement was soon signed with Whitbread and late in 1959 Rhymney swooped to capture troubled Ely Brewery, considerably strengthening the company and switching the power base firmly into Cardiff. Rhymney's head office was moved to the capital from Rhymney in 1961.

The takeover at the same time ensured Ely did not fall into 'unfriendly and competitive hands'. This was the start of the ruthless era when six giant brewing groups were formed in Britain. Rhymney was already spoken for. It was part of the Whitbread empire and the majority of the board were Whitbread men. The Ely takeover meant that, by 1961, Rhymney's pre-tax profits had leaped the half-million bar to £520,787. But expenditure was even higher, with £500,000 alone spent on a new brewery.

The Ely Brewery was demolished in 1963 after Crosswells had been considerably expanded in 1962. The additions took fourteen months to complete. Virtually a new brewery was built in Norbury Road with modern offices opposite on the site of the old Ely cattle market. They were officially opened in February 1963 by the chairman of the National Coal Board, Lord Robens, who was delighted with the new coal-fired boilers. The plant was able to brew 200,000 pints of draught beer a day – 'The Best Round Here' according to Rhymney's slogan – and handle 21,600 half-pints of bottled beer an hour, including Mackeson for Whitbread.

Rhymney moved into hotels in a big way, buying one of Cardiff's best-known landmarks, the Angel Hotel, in 1962. Other purchases followed in Porthcawl and Swansea, in co-operation with hotel group

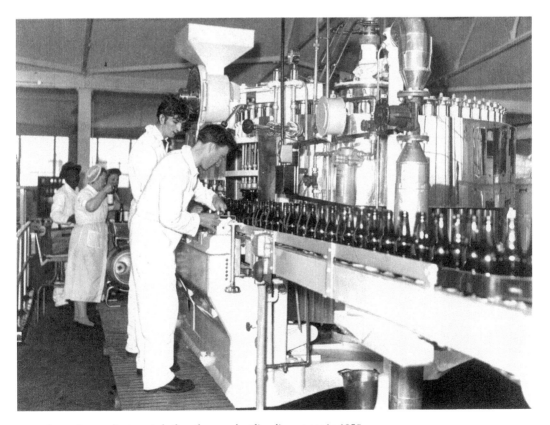

Last-minute adjustments before the new bottling line opens in 1958.

Forte's, and a motel was planned at Castleton, between Cardiff and Newport. Colonel Llewellyn was also pleased to sponsor the Rhymney steeplechase at Chepstow Races. But the hobby horse was having to gallop fast to stand still, as the new national groups competed hard for business. It was proving expensive to keep up. In 1964 the company issued 4 million shares to raise over £1.1 million.

Whitbread sensed it was time to take full control. In July 1965 Bill Whitbread arranged a meeting with Harry Llewellyn 'to discuss the financial situation'. Negotiations continued for six months, but there was only ever going to be one outcome. In January 1966 the inevitable happened. Whitbread bid £4.2 million for Rhymney, in which it already held a 30 per cent share stake. Equally inevitably, the Rhymney board, packed with Whitbread men, recommended acceptance. Not everyone was happy.

Arthur Smith, whose Newport Wine Lodge chain of off-licences had been bought by Rhymney in 1964, said, 'I wouldn't accept that for my shares.' The *Western Mail's* City Editor, John Roberts, commented that the offer was far from generous for a company estimated to be worth £7.75 million. The offer valued the shares, which had been as high as 9s 9d that year, at just 8s 2d. The offer does not 'inspire confidence in Whitbread's open-handedness,' said Roberts.

But Whitbread did not need to put its hand deep into its pocket, since it already held most of the cards. Even that seasoned warrior Julian Hodge, whose Hodge Group of companies held a substantial number of shares, recognised there was little point in haggling, even though his watchdog, Investors Protection Facilities Ltd, had received complaints from shareholders. Within a few days he had accepted the offer. The investment director of the Hodge Group, Alan Thomas, admitted, 'The acquisition price is poor in terms of asset values. But the price is probably fair when related to Rhymney's likely profit levels over the next few years.' Whitbread had got away with a bargain, on the promise of better dividends in the future. It had seized a group with two breweries and some 700 pubs. One shareholder described the deal as 'a marriage with every sign of a shotgun in the background'.

Colonel Bill Whitbread was on a roll through South Wales and the following year he bought Evan Evans Bevan's Vale of Neath Brewery in Neath with 240 pubs. In 1969 this was combined with Rhymney to form Whitbread Wales under the chairmanship of Harry Llewellyn. The hind's head rapidly replaced the hobby horse on pub signs, while beers like Whitbread Tankard and Trophy took over at the bar.

The brewery at Neath was shut in 1970 and Rhymney closed eight years later, but the Crosswells site, now known as Whitbread Cardiff or Ely, lingered on until 1982. It was only shut after Whitbread had completed a vast £52 million brewery in South Wales alongside the M4 at Magor in 1979, mainly devoted to brewing international lagers like Heineken and Stella Artois. Production of the Cardiff brewery's last beer with a local identity, the low-gravity keg Welsh Bitter, was transferred to Magor. The final brew at Cardiff was on 10 March 1982. The brewery was demolished for housing. The fate of the modern brewery office block might have amused old temperance campaigners. It was taken over in 1989 by South Wales Police.

But Whitbread did not pull out of brewing in Cardiff completely. A microbrewery was installed in a flagship estate pub it opened in 1983, the Heritage Inn in St Mellons, the first new pub it had built in Wales for ten years. Landlord Jack Madden was trained to brew Heritage Ale using malt extract, and customers could see the brewery through a glass screen. Old brewing vessels from Crosswells were even built into the pub 'to add an authentic touch to the interior décor', while coopering tools and hopsacks hung on the walls. But the brewery proved only a short-lived gimmick and was closed within three years.

Whitbread Wales disappeared in 1987 in one company reorganisation, being absorbed into Whitbread Trading South West, with its depot in Ipswich Road closing by 1994. It was turned

Whitbread's huge modern brewery at Magor. Once it opened, the Cardiff brewery's days were numbered.

into a David Lloyd leisure club, one of Whitbread's many new interests. It was turning its back on beer, despite producing a bottled celebration ale, Cwrw Dathlu Cynulliad Cymru, to mark the opening of the National Assembly in Cardiff in 1999. In 2000, it sold its breweries, including Magor, to the Belgian international brewing conglomerate, Interbrew. Where once Crosswells ales, then Rhymney's and Whitbread's beers had ruled the bars, now 'reassuringly expensive' Stella Artois poured supreme.

Bullmastiff

In the shadow of Cardiff's brewing giants, most of them now just ghosts haunting yesterday's pub cellars, there's one fresh pup barking away. Bullmastiff Brewery first cocked its leg against the established industry in 1987, when milkman Bob Jenkins turned his passion for brewing his own beer in his garage into a business. The eight-barrel plant was even brought in from Monmouth on three milk floats.

In partnership with his brother Paul, he set up his first brewery in an industrial unit in Penarth docks, and then moved in 1992 to a larger unit in Leckwith, Cardiff. Named after his prize pets, the brewery has proved top dog, winning a string of awards for the quality of its beers, with his Son of a Bitch strong ale taking a silver medal at the Great British Beer Festival in London in 1996, while Gold Brew was Champion Beer of Wales in both 1999 and 2000.

The premises off Hadfield Road may look familiar to fans of the S4C Welsh-language soap opera *Pobol y Cwm*, since two of the characters in the TV series set up their own Bragdy Cic

Mascots Kirsty and Mellie enjoy a glass of Bullmastiff's Ebony Dark with brewing brothers Bob and Paul Jenkins, soon after the brewery opened in 1987.

Mul (Kicking Mule Brewery). All the brewing action is shot at Bullmastiff. Literally. In one episode a car roared up to the brewery and a man leaped out with a shotgun and fired through the window.

Bob Jenkins prefers a quieter life, just brewing quality ales for around thirty customers across South Wales.

three

THE BEER

Old Owen by all was an oracle thought,
While they drank not a joke failed to hit;
But Owen, at last, by experience was taught,
That wisdom is better than wit.
One night his cot could scarce hold the gay rout,
The next not a soul heard his tale;
The moral is simply they'd fairly drank out
His barrel of old humming ale.

Charles Dibdin

If the past is another country, yesterday's beer poured from a different planet. Welsh ale was once famous and highly valued. Many early land agreements, listing payments in kind, mentioned Welsh ale. It was a distinct pre-Saxon drink, driven west by the invaders and then welcomed back into England as its heady flavour was appreciated by the new rulers. It was a strong, heavy brew, laced with exotic spices like cinnamon, cloves and ginger. It was the syrup of legends. Giants Gog and Magog were reputed to have swelled to their monstrous size on goblets of the liquor. It also oiled the machinery of early society. Denewulf, Bishop of Winchester, when leasing an estate from King Edward in the year 901, had to pay twelve sesters of 'sweet Welsh ale' as part of his annual rent. This was most likely flavoured with honey. Mead, fermented honey, was the most highly prized Celtic drink of all.

The spiced ale known as bragawd, bragot or braggett fell out of fashion when the hop hopped into Britain from the fourteenth century. But its reputation lingered long in the mind. When brewery chronicler Alfred Barnard visited Brain's in 1890, he could not resist asking head brewer, Mr G.J. Gard, who had been brewing in Cardiff for many years, about it. 'We wooed him cautiously to tell us how Welsh beer was made – that ambrosial liquor which was so dear to the hearts of our ancestors.' Mr Gard, who was keen to show his visitors his gleaming new brewhouse, cut his inquiries short. 'The old style of Welsh brewing has long since been abandoned,' he told

Young and old enjoy beer in Cardiff today
– but the brew has changed considerably
since a century ago.

him. But if bragot had disappeared, what was served over the bars in Victorian Cardiff would be
almost as strange to today's drinkers. There were not only few prominent brands, but beer was
viewed in a completely different way.

When beer had been produced at home or in the back of the inn, it was originally all about
'runnings'. The first run of water through the malt mash produced the strongest ale as it absorbed
the most sugars. The second running did not have as much body, while the third produced weak,
table beer. This heritage meant that beer tended to be considered in grades of strength or quality
rather than in different styles.

There was an additional divide between what was known as 'fresh' beer, which was brewed
and consumed within a few weeks, and old ales, which were matured for many months. These
'keeping' beers were the more powerful, expensive brews and the extended maturation not only
increased their strength, but often added a sharp touch of sourness to lift the heavy body of these
overpowering ales. Sometimes old ales were known as stale ales. Their 'off' flavours would not
be appreciated by many drinkers today, but were popular in Victorian times. Some liked to mix
these old ales with fresher milds in their drinking pots to create their own favourite flavours.
A few brewers specialised in these keeping beers. Hancock's Anchor Brewery in Newport
developed a reputation for its Hancock's Old Ale. 'Some local brewers took our old beer, and
against this we bought from them fresh beer to supply to our own houses,' recalled their first
salesman in the town, Mr J.A. Green.

The late nineteenth-century photograph (see opposite page) of the neighbouring Rock
and Fountain and Pineapple inns in St Mary Street, near the old Cardiff Town Hall, shows the
Pineapple Inn promoting 'Old and Mild Ales' above its entrance. These were the staple, local
drinks of the growing town. The mild would be the fresh beer, the basic running beer of the

Above: The heavier, sometimes sour, taste of beers in the past would not be to everyone's taste today.

Right: The Pineapple Inn in St Mary Street in the early 1890s advertises 'Old and Mild Ales'. Two doors down is the Rock and Fountain. Both houses soon vanished.

streets. Henry Anthony of the Castle Brewery in Great Frederick Street advertised himself in *Butcher's Cardiff Directory* in 1880 as simply a brewer of 'Celebrated Mild Ales'. Cardiff's commerce, coal docks and shipping were kept afloat on a sea of mild, popularly known as dark.

The Pineapple's windows also advertise Burton pale ales from Ind Coope. These hoppier Burton ales, notably from Bass and Worthington, complemented the local mild and old ales. Some local brewers also attempted to brew their own pale ales. An 1875 advertisement from James Ward's Tower Brewery in Ely shows the beers divided between five grades of 'Not Bitter' (i.e. mild), ranging from cheap XX at a shilling a gallon up to the most expensive old ale, XXXK OL, and three qualities of 'Bitter', including an India Pale Ale, the style popularised by Burton, originally for export to the Empire. The heavier hopping ensured these pale ales would keep longer.

The price list also shows the third main beer style of stout, with the Ely brewery offering a powerful imperial stout, as well as a double and a single stout. These differed in using darker, well-cured malts and sometimes roasted unmalted barley, and like old ales were matured in large vats. This again was a speciality brew, with much stout being imported into Cardiff, notably from Ireland and London.

London was also the capital of stout's weaker relative, porter. This style had risen out of the growth of Britain's first major commercial breweries in London in the eighteenth century. It was an attempt to produce a matured beer but on a mass scale and at a lower strength and cheaper price; a beer that boasted the characteristics of mixing older ales and mild beer, but in a single, entire brew. To do this, brewers like Truman and Whitbread built towering maturation vats. Some were so large, huge dinner parties were held inside to mark their completion, and when one vast vat at Meux's brewery off Tottenham Court Road, containing around 3,500 barrels of beer, collapsed under the pressure in 1814, many walls and buildings were swept away and several people drowned in the deluge.

Few of Wales's much smaller breweries could afford to invest on this scale, and porter never seems to have become a major drink in Cardiff like it did in London or Bristol. Frederick Prosser, when he ran the Old Brewery in 1860, did advertise himself as a beer and porter brewer. But by the time brewers like Samuel Brain built much larger plants in the late 1880s, the demand for porter was already in sharp decline, and Alfred Barnard's detailed description of his tour of Brain's grand brewhouse did not mention any large maturation vessels. They were notable by their absence.

Cardiff's brewers could barely keep up with the demand for fresh beer, and premium ales and stouts flowed in from outside to fill the gap and the empty tankards. This trade is best reflected in perhaps the most elegant pub mirror in the whole of Cardiff. It used to decorate the bar at the infamous Custom House at the top of Bute Street, but its rowdy reputation as a red-light resort saw the magnificent mirror moved to the Duke of Wellington in The Hayes, just behind Brain's brewery. Then Cardiff city centre's reputation as a wild night out saw it switched again to the suburban Heath Hotel in Whitchurch Road. All three were Brain's pubs, but the mirror did not advertise its beers. It carries the emblem of the Crown Brewery, which steamed away in John Street, opposite the Custom House, from 1874. Taken over by George Watson the following year, it became the Cardiff Brewery Company in 1885. But the beers proclaimed in etched letters and gold leaf were not brewed in Cardiff. Mr Watson's company also acted as agents for a number of prominent outside brewers, and it was their beers which were advertised in gilded highlights – Salt's Pale Ale from Burton upon-Trent, George's Porter from Bristol and Davis, Strangman's Irish Stout from Waterford.

The Cardiff Brewery Company was modest about its own beers, which do not even merit a mention on the huge multicoloured mirror. A book produced on *The Ports of the Bristol Channel* in 1893 reveals why. They were the town's bread-and-butter brews, much drunk but

Above left: Hoppier pale ales like Bass from Burton were popular in Cardiff.

Above right: Worthington of Burton upon Trent had close links with Hancock's. This festive advert appeared in the brewery's magazine in December 1926.

Right: The notorious Custom House in Bute Street, still standing in 1995 as a reminder of the wild nights 'below the bridge'. It was demolished three years later, along with the Glendower nearby and the Crown opposite.

Many pubs in Bute Street, like the Angelsey in 1907, advertised lager beer in their windows. It was said to be 'Swiss style'.

rarely promoted. A brief profile of the brewery declares, 'A considerable reputation has been gained by the productions of the company, chiefly mild ales.' But it admits, 'The place is taxed to its utmost capacity, and has become much too small for the requirements of the company.' Like most Cardiff breweries, it was struggling to meet demand, only being able to produce 100 barrels a week on its present plant. 'It is, therefore, their intention, as soon as suitable ground can be acquired, to build a new brewery, which will be on a greatly extended scale.' This expansion never happened, and the company, after being briefly taken over by its manager Francis Soule, was wound up. Brewing was a highly competitive market.

Even brewers from as far as Scotland poured into Wales, and not only famous names like Younger and McEwan. More obscure Edinburgh brewers also rolled their barrels all the way to Cardiff. The gable end of the Carpenters Arms in Newport Road, Rumney, prominently advertised 'Ritchie's Prize Medal Scotch Ale on draught' during the First World War.

Victorian Cardiff was awash with 'foreign' beer, and it wasn't just from Britain and Ireland. Some was from the Continent. For the busy docks were full of foreign sailors and they brought with them foreign tastes – including a liking for lager. Lager was almost unknown in the bars of Britain and the few British brewers who attempted to produce this beer found little demand for it at home. Ventures like the Wrexham Lager Company, which began brewing in 1883, had to rely on exports to stay in business. But cosmopolitan Cardiff was decades ahead of the market. Many of the pubs lining Bute Street, the long dive to the docks, advertised 'Lager Beer' in their windows and on their walls.

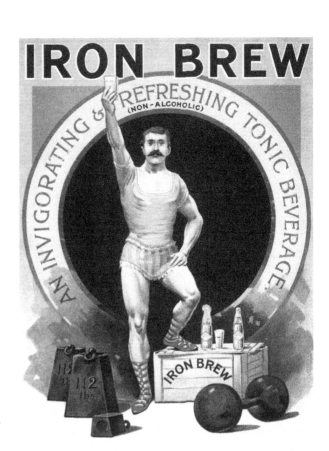

Non-alcoholic botanic brews liked to present a manly image. It didn't always work. This poster is from 1898.

Because of their lighter, drier character, lagers were sometimes sold as 'temperance' beers. But the temperance movement knew that if it wished to wean people off beer entirely, it would have to offer parched workers an alternative. So the campaigners supported botanical brewers, named after the various herbs they used. Some made no secret of their aim. Cox of London called their non-intoxicant ale 'Anti-Burton'. The herbal brewers produced brews varying from hop bitters and ginger beer to burdock ale and dandelion stout. Some had names designed to appeal to the macho working man, like Football Cup or Iron Brew. Howell Davies's Quaker Pep from Abercynon featured a miner on the label. The drinks were not supposed to contain more than 1 per cent alcohol, but could be quite intoxicating if the makers were heavy-handed with the sugar. One vendor prosecuted for selling ginger beer containing more than 5 per cent alcohol staggered magistrates by claiming men became sober on his brew.

Thomas Elliott of Gladstone Street boasted in 1893 that its speciality was 'improved fermented ginger beer' and it employed thirty horses and vans on daily deliveries. It was far from modest about its beverages. Its 'King of Tonics' Vilroy was claimed to be excellent 'for brain, nerve, muscle, vigour, digestion, stamina and perfect health'. Other botanical brewers in Cardiff before the First World War included Ridgway Brothers in Diamond Street, Spooner of Penywain Road, Lowe Brothers of Glamorgan Street, Clayton's New Era Beer of Penarth Road and Watson's Non-Alcoholic Brewery of Machen Place. Most supplied their botanical brews in large stoneware jars – which proved to be their Achilles heel. Percy Ridgway later recalled, 'This was their greatest expense, as with the ridiculously low deposit of a halfpenny or penny per jar, precious few ever found their way back to the factory, ending up as ideal containers for home-brewing of wine or even as hot water bottles.' Lowe Brothers of Pontcanna even branded their jars with the warning, 'Unauthorized persons using or detaining this jar will be prosecuted', but to no avail.

Most botanical brewers soon disappeared after the war or switched entirely to producing more conventional mineral waters and soft drinks, if they could afford the extra investment in bottling machinery. Not all were committed to temperance principles. William Bunning of Miskin Street, besides brewing stone ginger beer and hop ale, was also a beer, wine and spirit merchant, and bottler of Bass and Guinness in 1899. Some of Cardiff's brewers also kept a careful eye on this market in case they were forced out of brewing conventional ales by prohibition. William Hancock acted as agent for Kops Ale and Stout in Cardiff. This temperance brewery had been formed in London in 1892 with a huge capital of £250,000, and claimed to be selling 2 million bottles a week by the end of the nineteenth century. But most brewers jeered from the sidelines. In 1922 Buchan's Rhymney Brewery commissioned cartoonist H.M. Bateman to produce a cruel cartoon for its calendar, showing a group of unhappy miners at the bar, gazing into their glasses at being served 'Pansy Extract for Miners'. The brewery's advice was simple, 'Stick to Buchan's'. And most miners did. They preferred beer with a comforting alcoholic kick.

Frank Thatcher, the head brewer of Marston's of Burton upon Trent, wrote a *Treatise of Practical Brewing and Malting* in 1905. In the weighty volume, he sketched out a beer map of Britain, with Burton noted for its pale and strong ales, London for mild and sweeter stouts, Edinburgh for special Scotch ales, and Dublin and Cork for Irish porter and dry stout. Cardiff did not get a mention, but it was clear where it belonged. 'In districts where miners consume the beers produced, it is usual to aim at a luscious palate fullness and a sweet type of mild ale.' These were brewed with 'a large percentage of sugar in the copper and well-cured malts in the mash tun'. They were not only refreshing, but with their fuller flavour also satisfyingly filling and quite intoxicating. They were not as heavily soporific as the vatted old ales, but still stronger and much more warming than the standard bitter today. In 1880, Gladstone reckoned that the average original gravity of mild ale was 1057, around 5.5 per cent alcohol. Mining was a hard occupation and the men wanted more than just a thirst-quencher. They needed something nourishing and sustaining. They viewed beer, made from grain, as liquid bread. Mild, at a popular price, was their preferred choice. And just as coal cascaded into Cardiff from the mining valleys, so the deep thirsts from the pits were also adopted in the dusty docks.

But many workers were also wary of the new breed of commercial brewers, with their more scientific equipment, who had sprung up to replace publican brewers. They feared they might be adulterating or weakening their natural beer. Before, if they had not been satisfied with their pint, they could berate the man responsible directly. He was just the other side of the bar. Now the landlord could blame a distant brewer. The temperance movement was not slow to exploit this alienation.

MP Sir Thomas Whittaker, in a 1908 booklet for the Temperance Legislation League, attacked the brewers:

They talk about 'good malt liquor', but year by year they use more and more sugar, syrups, glucose, saccharin and other preparations and concoctions. The use of these cheap substitutes has increased tenfold. In addition, there has been an enormous growth in the use of rice and rice grits, maize grits, flaked rice and the like. During the last ten years the use of this latter class of substitutes has nearly doubled. The proportion of malt has gone steadily down, and the proportion of substitutes and preparations has steadily gone up.

It was not an assessment commercial brewers could deny. Frank Thatcher, in his recipe for mild, not only recommended the use of 18-20 per cent of glucose or invert sugar but also 10-15 per cent of prepared rice or maize. Many brewers used Beanes' Patent Grist produced from rice which, added to the mash tun, was said to be able to stop the brew going sour and ensure the brilliancy of the beer. Beanes, a London company, even remarkably claimed that 'beers brewed with this material will remain sound for years, even if of low original gravity'.

Beanes was one of many companies exploiting a growing problem in the brewhouse. Preserving the powerful vatted ales had rarely been an issue because of their high alcohol content. The lower-gravity milds presented more difficulties, as the strength of these brews continued to slip in the face of the sharply rising cost of buying pubs, the demands of shareholders for fat dividends and, particularly in South Wales, the clamour of the temperance movement. E.A. Pratt in his book *The Licensed Trade* of 1907 said, 'Comparing the ordinary beers now being consumed with those of the period in question (twenty-five years ago) one finds that in actual alcoholic strength there has been a reduction of between 15 and 20 per cent.' Ensuring the condition of the beer was becoming more of a problem as brewers took over more and more pubs and extended their supply lines. The increasing use of glass, both in bottled beers and drinking vessels, also exposed the brewer's art to critical drinkers' eyes.

Whittaker pressed home his attack under the heading 'Brewery or Chemical Works?' He wrote:

In the report of the Beer Materials Committee published in 1899, there is a formidable list of chemicals, colouring matters, headings, primings, preservatives, substitutes and sundries, which Mr Richard Bannister of the Inland Revenue informed the committee were advertised in trade papers for brewers' use. Now we begin to see how the brewer and the publican have been taking it out of the working man for years past. They have not given him the benefit of reduced taxation, cheaper materials and improved methods. On the contrary their advocate, Mr Pratt, admits that they have steadily reduced the strength of his beer, year by year; and I have shown that, increasingly during recent years, they have substituted all kinds of things for genuine malt. They also doctor the liquor with innumerable chemicals and concoctions which suggest chemical works rather than a brewery.

The brewers were vulnerable to this line of attack after a sensational incident late in 1900. Three double-column headlines dominated the *Manchester Evening Chronicle* on Friday 23 November: 'Poisoning in Manchester – Remarkable Revelations – Arsenic in Beer'. A Salford workhouse physician, Dr Ernest Reynolds, had linked unusual illnesses to beer drinking. When he tested local ales he found arsenic 'in dangerous proportions'. When he published his findings in the *British Medical Journal*, it triggered uproar, outrage and panic across the country. The *Manchester Evening News* called it 'Almost a Plague'. Soon suspect stocks of beer were found across north-west England, with Liverpool badly hit. Thousands were ill and hundreds died.

Brewers at first flatly denied their beer was to blame. They pointed the finger at impure water supplies or fever brought back from Africa by Boer War soldiers. Others blamed unscrupulous landlords, chemicals used to clean casks or sulphur dressings on hops to control pests. The angry hop growers hit back, claiming brewers added phosphoric acid to their old ales to reduce maturation time. Beer's healthy reputation was being shredded. Temperance activists in Cardiff used the panic to cast doubts on the safety of Welsh ales. Sales slumped.

Public health officials eventually tracked down the guilty party – sugar manufacturer Bostock's of Garston, Liverpool, whose glucose and invert sugar contained arsenious oxide due to the 'exceptionally high' levels of arsenic in the sulphuric acid used to make them. But if the brewers

Scottish brewer McEwans used the arsenic scare in 1900 to sell its beer as 'absolutely free from impurities'.

thought they could now get back to business as usual, once dangerous sugar stocks had been destroyed and contaminated beer recalled, they were in for a rude awakening. The dead might be buried but they could not be so easily forgotten.

In February 1901, the Government appointed a Royal Commission to investigate the mass poisoning. Its deliberations took almost three years, and put the whole industry and its brewing methods under the microscope for month after month. The incident also breathed fresh life into the pure beer campaign, which until now had largely been an agricultural lobby, trying to persuade Britain's brewers to use more British barley, rather than foreign cane sugar. A Beer Bill proposed dividing British beer into all-malt and part-malt brews. But support drained away once the commission discovered some malt samples also contained low levels of arsenic from the gas coke used in some maltings. When the Commission finally reported late in 1903, its recommendations mainly concerned purity checks and controls. But many brewers stopped using sugars and other adjuncts in large amounts. Some found it paid to promote their beers as brewed only from malt and hops. Chemicals also belatedly came in for much closer scrutiny.

The industry had been shaken out of its complacency. Many drinkers had for a while switched to wines or spirits. Beer had come close to losing its reputation. The trade was still sensitive about the issue in 1928, when Hancock's repeated an extract from the *Brewers' Guardian* in its house journal:

During the recent visit of the British Medical Association to Cardiff, the members inspected the brewery of Messrs Hancock & Co. Ltd. We are sure this did an enormous amount of good, as the medical gentlemen were able to see the care and cleanliness necessarily displayed in modern brewing. They could also see for themselves that beer is not composed of a mass of chemicals as our detractors allege.

Hancock's John Bull figure was enlisted to the cause of promoting beer's healthy properties.

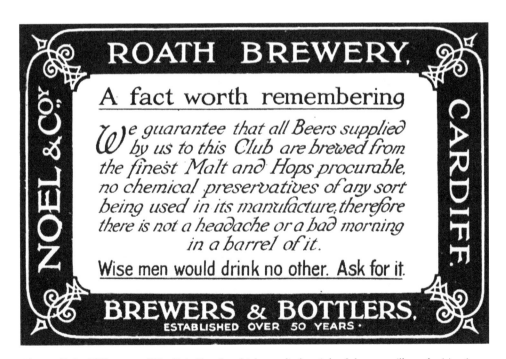

The small Cardiff brewer of Noel's in Roath, which supplied mainly clubs, was still emphasising in the 1930s that its beer contained no chemicals: 'There is not a headache or a bad morning in a barrel of it'. The brewery closed in 1956.

One notorious cartoon in the *Daily Dispatch* in 1900, with the title 'Beer, Glorious Beer!', had shown a brewer's assistant handling sacks labelled salt, sugar and sulphuric acid, 'compound essence of factory sweepings' and arsenic, and 'other concocted items'. 'What about these 'ere, guv'nor?' he asks the brewer. 'Chuck 'em all in, I'm only mixin' a fourp'ny,' he replies. Fourpenny (the price of a quart) was one name for mild.

It's difficult today to realise just how far mild once dominated the market. In the late 1930s, the *Brewers' Journal* estimated it accounted for more than three-quarters of all beer brewed in Britain. In industrial South Wales that figure was certainly higher, with some breweries producing both light and dark milds. And this popularity remained strong despite the fact that mild's strength had been slashed during the First World War. Before the conflict, the average original gravity of all beer had been 1052 (around 5 per cent alcohol). By 1918 it had plunged to 1030 (3 per cent) and though it quickly recovered to just above 1040 (4 per cent) it never returned to the more powerful levels of the Edwardian era. It was part of the industry's unwritten pact with the Government that, to avoid prohibition, beer would be less intoxicating.

But in South Wales, the brewers had already bent before the blasts of the temperance preachers long before the restrictions of the First World War. In 1894, when the average original gravity of mild across the UK was above 1050 (more than 5 per cent alcohol), Hancock's popular XX was 1043. By 1905 it had fallen to 1040 (4 per cent alcohol). It remained at this level until 1915, before the death of beer's body in the First World War.

An advert for Hancock's Dark Malt, top left, high on the wall of the Red Lion on the corner of Kingsway and Queen Street between the wars.

Ken Morison of Welsh Brewers, in the Cardiff sample cellar in 1978 with quality control manager Mike Jones, left.

That year the National Defence League, set up to protect the industry against a hostile government and determined dry lobby, produced a table of beer strengths for its No. 12 district, which covered Cardiff and the mining valleys. Twelve of the twenty beers featured were classed as XX or fresh beer. It showed that local rivals Crosswells (1042) and Ely (1043.4) were slightly stronger than Hancock's (1040), while breweries up the valleys, like Treherbert and Giles & Harrap of Merthyr Tydfil (both 1038) tended to be weaker. Pontycapel, once noted for its strong ales, brewed an XX mild at 1032. The strongest was from the Pentre Brewery at 1044.6. These contrast sharply with the more powerful pale ales from Burton brewers, like Allsopp and Worthington, whose brews range from 1050 to 1060, while Guinness's Extra Stout from Dublin weighed in at 1074 (more than 7 per cent alcohol).

By the end of the war the quality of XX had been so degraded that most brewers dropped the term as they struggled to revive sales. Ely decided to add an X and its basic dark mild became XXX. Hancock's further increased the X factor and produced a stronger XXXX (1051) with a higher hopping rate, besides a much weaker XPA (1032) and its popular light mild PA (1041). But it soon dropped XXXX in favour of a weaker Dark Malt (1042), with half the hops.

A company like Hancock's, which traded across much of South Wales, had to produce a variety of milds because local tastes differed from town to town, and valley to valley. Just like Frank Thatcher could draw up a beer map of Britain, so brewer Ken Morison, who joined Hancock's in 1940 before retiring as managing director of Welsh Brewers forty-four years later in 1984, could plot the preferences of each local district across his patch. 'The Aberdare Valley does not like very bitter beer. You would never be able to sell a dark, sweet beer in Newport or Merthyr, but it would probably go down well in parts of West Wales.' He spoke with authority because beer ran in his blood. His stepfather had been head brewer at Hancock's before him.

As for Cardiff, it was always firmly in the dark, as demonstrated by the brewery which branded its name across the heart of the city – Brain's.

The unique flavour of Cardiff
Pop along to your local Brains pub and get a taste of it!

Left: A 1980 advert emphasises that Brain's is soaked into the fabric of the city.

Opposite: Almost every pint sold in the city in the 1950s seemed to be Brain's Dark, with the odd bottle of Strong for mixing.

It's Brain's You Want

Verses in honour of the capital city always manage to rhyme Cardiff Arms Park with Brain's Dark. It was the first pint after work, the first, second and seventh pints after the rugby match and the prime lubrication for male voice choirs. It dominated the drinking culture of the city for decades. It was soaked into the streets. As a regular of forty years at the Packet, one of the few traditional Brain's pubs still standing in the docks, recalled, 'Everyone drank Dark or Dark and Bitter mixed. Men used to come in off the dock, covered in coal dust, for a pint.' Though its official title was Red Dragon, it was always known in Cardiff as Dark. Eventually the brewery also adopted this title. Brain's Bitter was often called Light.

Until the 1970s, Dark was the brewery's best-selling ale by far. It sold much more than all other Brain's beers combined. It was drunk by young and old, even hot or cold. David Jones, seventy-nine, of Splott, recalled in 1999 that he had his first taste of Dark at the age of seven. 'My grandmother used to send me out when I was young to get a jug filled up with ale. She would then mull it with a hot poker and give me a sip. But the sips just got bigger and bigger.' Some Cardiff pubs, like the Railway Inn in Fairwater, sold only Dark on draught until 1980.

For proof take a stroll up the metal stairs to the upper floor of the company's flagship pub, the Yard in St Mary Street, opened in 2003 on the site of the old brewery yard and the former brewery tap, the Albert. Above the arch where drays carrying barrel after barrel of Dark used to roar out into the traffic are glass cases containing a collection of Brain's memorabilia. Among them at times is an old handwritten stock ledger. When I last saw it, it was open at the entry for 30 September 1940. Then, out of a total stock of 1,579 barrels at the brewery, RD (Red Dragon) accounted for 1,155 barrels – almost three-quarters of the company's production. The next best seller was IPA (later replaced by SA) at just 213 barrels, followed by LB (Light Bitter) on 138.

This mighty mild trade was probably not what Samuel Brain had intended when he built his new brewhouse in 1887. When Alfred Barnard visited soon afterwards, he commented, 'Messrs Brain & Co. have now become so proficient in the art of brewing pale ales that the Cardiff publicans are not obliged to go to the "beer city" for their Burton ales, that commodity now being brewed on the same principle in their own town.' Samuel Brain was hoping to steal the invading brewers' more profitable business in premium beers. Barnard approved. 'We stepped inside [the sample room] and sampled two or three sorts – pale ale, mild beer and stout. The first we found to be a delicious, full-flavoured tonic beverage; the latter a nutritious, well-bodied drink, quite equal to the brews of London and Dublin.' The mild passed his lips without comment. But it was the beer Cardiff's workers demanded. Despite Samuel Brain's wishes, Dark dominated his trade. Dark so overshadowed the brewery that its official name, Red Dragon, was adopted as the brewery trademark.

Though its strength has been sapped by the rigours of two world wars, the pump-clip claims that it is still 'brewed to the original recipe'. In 1991, brewer Peter Brennan said that the strength (3.5 per cent alcohol) and recipe had remained unchanged for at least twenty-five years, as long as he could recall. 'It's a fine blend of English pale ale and roasted chocolate malts, Goldings and Fuggles hops from Worcestershire – and Welsh water,' enthused brewing director John Glazzard at the time, waxing lyrical about its dark ruby colour, pleasing aroma and tight creamy head. 'There should be rings of lacing down the glass, showing how many gulps the drinker has taken.' Unlike most other dark milds, which are sweeter and can be cloying, Brain's Dark is distinctively dry. It has a uniquely refreshing, roasted flavour. 'I can say quite objectively that Dark is on its own,' added John Glazzard. It was the toast and taste of the capital, but not so appreciated up the Valleys or to the west, where drinkers preferred a sweeter, maltier dark mild. Brain's Dark, surprisingly, has the same units of bitterness as the brewery's Bitter. Unlike its local rivals,

Above left: This old ashtray shows Brain's was once known for its bottled Imperial Stout, IPA and Home Brew as well as Red Dragon.

Above right: Home Brewed traded on fond memories of the old home-brew pubs.

Hancock's Dark Malt or Ely XXX, which sold across a much wider area, it was the defining beer of Cardiff. It was the Dark spirit of the city captured in a glass.

But Brain's brewing history is not completely absorbed in the Dark ages. An old ashtray betrays an even blacker past. The brewery was once noted for its Imperial Stout, or 'Little Imp', sold in nip-size bottles. The empire was not the British Empire – but the Russian. Extra-strong Russian or Imperial Stout had originally been brewed in London for export to the Baltic in the eighteenth century. This rich, intense brew, with a liquid fruit-cake character, had reputedly been a favourite of Catherine the Great, hence the Russian Imperial title. Barclay's of Southwark was the main producer of these vintage brews, which could be matured for more than a year in the brewery. But Cardiff, with its strong shipping connections, also for a while became an Imperial stronghold. Brain's was one of the few breweries to successfully copy this specialist style. It was the Old Brewery's tribute to the old vatted brews.

The other beer mentioned round the rim of the ashtray, besides draught Red Dragon and India Pale Ale, is Home Brew. Many breweries in South Wales and the West of England produced a bottled beer under this title, and the name reflects the persistence of home-brew pubs in the region, particularly in rural areas. In Abergavenny, for instance, six pubs were still brewing as late as 1934. Those that survived usually had a good reputation for their beer, since poor publican brewers soon went out of business or out of brewing. Commercial brewers, anxious to tap into this barrel of goodwill, produced a bottled Home Brew. Most of the main Cardiff brewers marketed one. Later Home Brew often became little more than another name for a basic brown ale, but originally it was a stronger, richer beer, a double brown. Its quality is shown in a 1934 price list. Brain's sold Home Brewed in both nip-size bottles and half-pints, but not in pints or flagons. It was the most expensive bottled brew along with IPA, costing 5d a half-pint. In contrast draught Red Dragon cost just a penny more, 6d, for a whole pint, while Brain's also produced a bottled Nut Brown Ale at 3d a half-pint.

By then Little Imp was no more than a distant hangover, as many of the really powerful brews had failed to survive the First World War. Brain's Home Brew seems to have been a casualty of the second global conflict. But most of the rest of the bottled beers on the 1934 price list would still be familiar to many drinkers today – Pale Ale, Extra Stout and Strong Ale. What would surprise is the large number of beers sold in two-pint flagons. Before canned beer and six-packs, these were the Sunday sustainers, keeping drinkers going through the dry Sabbath. Brain's served up five different beers in flagons – Ale and Mild Ale at the very cheap price of 10d for two pints, Bitter Ale at a shilling, and Strong Ale and Stout at 1s 2d.

The revered SA, named after Samuel Arthur Brain's first initials, did not appear on the bar until the mid-1950s, with beermats having to explain it was the brewery's best bitter. It later gained the alternative title of Skull Attack, but never deserved this reputation. It only seemed strong, at a modest 4.2 per cent alcohol, to drinkers brought up on a solid diet of Dark. SA was not the brewery's first attempt at a best bitter after the war. In 1949 it had launched Extra as 'the better bitter on draught' but it did not last long.

A price list for 1959 includes two other draught options for Brain's customers, though both at the same price as Red Dragon and Bitter (1s 1d a pint). One was MA, a special darker version of the bitter almost exclusively supplied to the Crown at Skewen, near Neath. The other was 'Half Red Dragon and Half Bitter'. The fact that this mix appeared on the price list shows how often it was requested. Changes in the bottled beer range included the replacement of Home Brewed by a cheaper Brown Ale at 9d, just a penny a half-pint more than Nut Brown.

The bottled beers were all brewed and filled at Brain's New Brewery in Roath, opened in 1919 to meet the growing demand for bottled beer. The site bottled Bass in an area known as the Bass Cellar, and handled other beers and cider. From 1932, the company also began to produce

Brain's SA has a bigger reputation at the bar, but traditionally most drinkers like Christopher Brain have preferred Dark.

its own soft drinks under the Spa trade mark. At the peak of demand for bottled beer in the 1950s, three bottling lines were in constant use.

But the company always kept faith in its traditional cask beers from the original Old Brewery in the centre of town. While many regional breweries rushed to compete with the national brewers' heavily advertised keg brands, launched in the 1960s, Brain's experienced a surge in demand for its three real ales, Dark, Bitter and SA. Only in the 1970s did it dip its toe into the keg market, first with Gold Dragon and Tudor Light, unsuccessfully promoted by a dopey-looking dragon, and replaced by Capital Keg in 1981.

In 1986 Brain's also started to can its beers and introduced large two-litre PET plastic bottles for the take-home trade in place of the famous flagons. It meant the end of an old joke. A football fan at a Cardiff City-Swansea derby match at Ninian Park consoles his neighbour, as the big glass bottles fly over their heads. 'Don't worry. You know what they say. It won't hit you, if it hasn't got your name on it.' The other man ducks. 'That's the trouble. I'm Christopher Brain.' The new PET bottles were not just for the low-gravity beers. In fact the first one was their strongest ale, IPA (1046), previously only available in small bottles, sold as Old Brewery Special Bitter.

But the really flying beer business was elsewhere, with 40 per cent of the British trade having switched to lager by 1985. Brain's ventures into this highly competitive market lasted little longer than its keg ales, though it globetrotted with enthusiasm to try and find the right brand. It first started looking for its own lager in 1982 and eventually chose Faust, launched in 1985. It was brewed under licence from the Faust family brewery of Miltenberg in Bavaria, with the yeast regularly flown in from Germany. It was a costly adventure. More than £1.5 million was spent on converting the New Brewery in Roath to lager production, including installing eight conical fermenting vessels. Brain's hoped this would open up a whole new market, but Faust, meaning fist in German, failed to punch its weight. It could not compete with the heavily advertised standard brands and was withdrawn by 1991.

Brain's then concentrated on its premium all-malt lager, Sternbrau, brewed under licence from Hurlimann of Zurich since 1986, with yeast flown in from Switzerland. Traditional German Hallertau hops were used. Brewer Paul Walkey kept a close eye on the beer, which was lagered (stored) for at least six weeks. A wooden cabinet in his office at the New Brewery opened to reveal two taps for dispensing Sternbrau. 'Only strictly for sampling, of course. We have to check the quality.' But with a trading agreement with major brewer Whitbread in 1990, and its lagers Heineken and Stella Artois on the bar, Hurlimann's star was quickly eclipsed. By 1993 the New Brewery and its lager plant and bottling lines were closed.

Exports were no more reliable. The first in 1988 was a premium bottled bitter to Wales's Celtic neighbour Brittany, with the label boasting its Welsh origins in both Breton – 'Bier bro Gembre' – and French, 'Bière du Pays de Galles'. Other forays into Europe included a stronger draught Celtic Dark to northern Italy in 1997. This followed Brain's first trip across the Atlantic in 1996, with a bottled and then keg version of SA sold as Traditional Welsh Ale.

At home a new generation of nitrogenated 'smooth' keg ales appeared in 1996, based on the existing Dark and Bitter, and these proved more successful than the earlier keg beers. Dark Smooth was originally introduced at a higher strength of 4 per cent alcohol, but soon dropped back to 3.5 per cent. A premium smooth ale, Dylan's (4.5 per cent), was launched with much fanfare in 1998, named after the poet Dylan Thomas, with high hopes that it would appeal to the American market. But after an early surge of sales it went 'gentle into that good night'.

But traditional ales still remained the backbone of the brewery, with additional beers like Buckley's Best Bitter and Reverend James also brewed in Cardiff following the takeover of Crown Buckley in 1997. The brewery had also from 1995 begun to surprise drinkers with a series of seasonal and special ales on the bar, such as Easter and Summer ales and Victory Ale to

Above: Victory Ale in 1995 was one of the first of Brain's special brews. Marketing managers Louise Prynne and David Knowles lead the celebrations.

Right: Brain's modern Yard bar in St Mary Street, with its tall stainless steel fonts, is now the best-selling outlet for cask Dark.

HANCOCKS BEERS

The Sign of Hospitality

Hancock's liked to present an olde worlde image, as in this 1950 poster.

mark the fiftieth anniversary of the end of the Second World War. Other beers like Arms Park Ale (Cardiff Rugby Club, 1999), Hat Trick (Glamorgan County Cricket Club, 2001) and Bread of Heaven (Wales rugby team, 2004) celebrated Brain's growing array of sponsorship deals. These one-off brews were relaunched in 2001 as A1 Ales, starting with Fresh Hop in the spring and including other former Buckley beers such as St David's Ale and Merlin's Oak.

But there had been a much greater and far more significant upheaval in Brain's regular ales. Dark had become a shadow of its former self. From being the dominant beer at the brewery until the early 1980s, it had been eclipsed by the Light. Brain's Bitter now ruled the bars, selling nearly three times as much. In 1991 Dark accounted for 20 per cent of production, compared to 57 per cent for Bitter. It was only just ahead of SA at 17 per cent. Ten years later it was down to 10 per cent.

But there was light at the end of the Dark downward tunnel. As Dark ceased to be the bread-and-butter beer of the brewery, so it was transformed into the brewery's speciality ale; a distinctive brew not found elsewhere. Drinkers across the UK began to discover it was a black beer with hidden depths. In 2001 it was judged Champion Mild of Britain at CAMRA's Great British Beer Festival in London, before taking the third place bronze in the overall Supreme Champion Beer of Britain. It may be low on alcohol, but it has unexpected strengths, with more taste than beers twice as strong.

The Yard, the bold £1.5 million venue which fronts the Old Brewery Quarter redevelopment on St Mary Street, is a far cry from the pubs where Dark once ruled the sawdust and spittoons. But it is where Dark shines in the twenty-first century. It has become Brain's best-selling outlet for the mild in Cardiff, helped by a little drama at the stainless steel bar. The tall spouts behind

PERFECTION
IN A NUT-SHELL

NUT BROWN 5^D
ALE PER BOTTLE

Nut Brown was launched in 1928 at
the premium price of 5d a bottle.

the steel handpumps are at eye-level. So instead of holding the glass down below the bar, staff lift the pint up, allowing customers a clear view of the glass being filled. It's beer reverence with a vengeance, as everyone gazes at the ale on high. As the brewery likes to boast, Dark is now the second best selling black beer in Wales after Guinness.

Have a Hancock's

Of all the Cardiff brewers, Hancock's most liked to present an olde worlde image. Its inn signs carried the claim, 'Hancock's – The Sign of Hospitality'. With its jovial John Bull figure and smart show teams of grey horses, it at times traded on a glowing, sepia-tinted view of the past. It liked to be seen as a reassuringly unchanging company in a rapidly changing world. But its beers were fluid. Behind the façade of solid tradition, the ales were always altering.

This is true of all breweries, but we know more facts about Hancock's because, after the Bass Charrington takeover, Norman Bridgwater, the quality control manager for Bass Production Wales, produced a technical report on the brewery in 1972. It detailed how the beer styles and strengths in Cardiff had been transformed down the decades. In 1894, when Hancock's took over the County Brewery off Penarth Road, it produced three milds – a cheap dark XX (1043), a mid-range XPA (1046) and a PA (1051). These strengths were low for the era and indicate both the influence of the Welsh temperance movement and the pressure to produce large amounts of fresh beer for the boom town. The really strong beers included an Extra Stout at around 1065 gravity, though this was still weaker than Guinness at the time. Edwardian price lists show the

Above left: Hancock's tried to brighten the post-war gloom in 1947 with a cartoon character, but many of their beers had lost their strength.

Above right: Five Five helped to restore the reputation of Hancock's bottled beer in the 1950s, once they had settled on a definitive label design.

most expensive beers were a matured old ale XXXX (which later became the name of one of Hancock's weakest milds!) and an Oatmeal Stout. There was also a Burton-style IPA.

After the trials and tribulations of the First World War, XX vanished and XPA became the brewery's cheapest ale with a gravity of just 1032 (around 3 per cent alcohol) with PA, a light mild, at 1040 (4 per cent). Dark Malt (1042) was Hancock's rival in Cardiff to Brain's Red Dragon. The famous HB, Hancock's best-known bitter, had also appeared by 1925, at a gravity of 1048 (almost 5 per cent), with a noticeably much higher hopping rate than the milds. Bottled beer sales were also starting to boom, with Hancock's, like Brain's, doing a thriving trade in two-pint flagons. A 1927 Christmas promotion offered to deliver flagons of five different brews to your door – Bitter Ale (10d), Amber Ale (1s), Strong Ale (1s 4d), Special Stout (1s 2d) and Oatmeal Stout (1s 4d). Then came Nut Brown.

In recent decades beers carrying this familiar name have often been poorly regarded weak brown ales, but when Hancock's launched Nut Brown in 1928, it was viewed as a premium ale. It was one of the strongest beers at the brewery, with a gravity of 1048 (Amber Ale and Stout were both 1041). Its high prestige is shown by the way it was rolled out. Instead of being slipped onto pub shelves, it was unveiled at Cardiff Races at Ely on the Easter bank holiday, when hundreds of dozens of bottles were supplied to racegoers. The brewery described it as 'a full-bodied, old-style type of ale' with 'a rich, dark colour'. It was Hancock's answer to Brain's

Home Brewed, selling at the same price, 5d a half-pint. At times and in different trading areas, it was sold as Home Brewed and, just to add to the confusion, Hancock's HB Bitter was also known as Home Brewed.

Nut Brown's sales were as sweet as a nut. When Hancock's produced a booklet in 1937 to celebrate the first fifty years of the company, it was one of only two Hancock's beers mentioned by name in the 108 pages:

> Hancock's buyers select only barley that has ripened with a good allowance of sunshine and has thus become perfect grain from which to brew the clear, golden light ales for which the brewery is famous, while the same grain is also best for all other brews, including Hancock's Nut Brown Ale, long a favourite in the Principality.

The description reveals that Hancock's reputation, unlike Brain's, rested more on its light than its dark milds. The other beer highlighted was Amber Ale, in a report on the fermentation room where the wort (the liquid extract of the malt) is mixed with yeast 'to become pale ale, Amber Ale, creamy stout or one of the many other brews which have made Hancock's ales famous.'

But two other beers were prominently featured in the booklet, each being promoted with full-page adverts. The foreign sailors in Cardiff docks were still thirsty for a taste of home and Hancock's had cornered the local demand for lager. The brewery was the sole agent in South Wales for two brands. Neither will be recognised by lager drinkers today. One was Patzenhofer from Germany. 'It is real lager. Those who have once savoured the true lager qualities of "Patz" will have no other.' It claimed to hold the world-record sale for lager (88 million gallons) and to account for half of all German beer sold in Britain. It was matured for twelve months. The other was Barclay's lager from London, which was available in both light and dark forms. Both were sold as upmarket drinks by well-dressed men and women. The Barclay's couple were waiting to board a biplane. There wasn't a miner in sight. Lager was still a niche business in Britain. It was small but elegant beer compared to mild. Lager louts were lurking a long way in the future.

What was not far off was the Second World War, which not only destroyed Patzenhofer's hopes of global domination, but also torpedoed the surviving strength of British beer below the water line. The gravity of Hancock's Dark Malt plunged from 1042 before the war to 1034 by 1945. But unlike the First World War, beer never bounced back. Rationing and post-war restrictions reduced Dark Malt further to 1032 by 1955. The popular light mild, PA, also fell to the same low level. While two other Hancock's milds in 1955, XXXX (1031) and XPA (1030) were even weaker. Draught mild's reputation never fully recovered. It was increasingly viewed as a wishy-washy beer and some drinkers began to switch to bitter. Over the same period, Hancock's HB had only fallen to 1039 in 1955. The two weakest milds, XXXX and XPA, were withdrawn by 1960.

With draught mild more difficult to keep in good condition at these low alcohol levels of around 3 per cent, other drinkers turned to more reliable, pasteurised, bottled beer. Only to find many had lost their bottle as well. Nut Brown Ale, launched with such a fanfare in 1928, had sunk from 1048 to 1033 by 1960. While Hancock's Strong Ale defied the Trades Descriptions Act, with a gravity of 1030. In 1960 bottled Strong Ale was the brewery's weakest beer. It was only strong on deception. It is easy to see why the brewers' organisations fought for so long to keep the strength of their beers a secret from the public.

Behind the closed brewhouse door another major change had happened after the Second World War. Until the 1950s Hancock's had, unusually in Cardiff, used the dropping system of fermentation. Instead of the wort being fermented in one vessel for a week, with excess yeast skimmed off, it was fermented in a top vessel and then, after three days, run into a shallower vessel on the floor below, stimulating the yeast. This more intricate system, widely used in Scotland, helped to produce cleaner, well-aerated, fully fermented beers, with unwanted residual matter left behind in the top vessel. It was ideal for the lower-gravity light milds for which Hancock's was famous. It was the system which created Hancock's character. But it did require double the amount of vessels and was more expensive to maintain. In the 1950s the company quietly abandoned this system, doubling its fermentation capacity from 2,000 to 4,000 barrels a week.

Hancock's venture into the export market after the war, with stronger bottled beers, did give rise to a genuinely strong ale, originally with a gravity of 1055, though it was soon lowered to 1048. But the company had difficulty getting drinkers to ask for it. The problem was the name, Five Five. When first introduced, the label also carried the number 55 in red. So customers asked for Fifty Five. Hancock's added a dash between the figures with little success, and then abandoned the numbers altogether, just spelling out Five Five. There was also briefly a Seven Seven export stout, but Five Five lasted longer. It became the company's new flagship beer in distinctive, dumpy bottles – along with a processed mild.

Hancock's had long been interested in new ways of keeping and serving beer. As early as 1906, it had offered pubs 'pressure beer' for which it charged a shilling extra a barrel, including fixing the necessary 'apparatus'. This early experiment failed to froth, but when pasteurised and pressurised keg beer started to take off, Hancock's was keen to exploit the market. As early as 1960, two years before a major brand like Double Diamond, it introduced Barleybrite, a filtered and pasteurised version of its HB Bitter, served from a plastic wheatsheaf on the bar. Unusually it used nitrogen like Guinness, rather than carbon dioxide, to dispense the beer, giving a creamy texture.

Keg beers were at first seen as a premium product with high initial costs – and were priced accordingly. But this was not a recipe for success in South Wales, where low-gravity, cheap milds were still immensely popular. In 1963, Hancock's introduced a filtered tank beer, Superex, based on its popular light mild PA, delivered by road tankers and then piped into five-barrel cellar tanks in the pub or club. This bulky system was initially very successful, but proved inflexible as tastes broadened and was overtaken by Allbright, a keg version of PA introduced the following year.

After the takeover in 1968 by Bass Charrington, most Hancock's beers like Five Five, Amber Ale, Barleybrite and Superex quickly disappeared. The John Bull figure only still lifted his glass on pump-clips for Hancock's HB and, for a while, PA. The brewery was now part of the Bass empire and the group's beers took over. Dark Malt was replaced by Worthington Dark, but the new dark brew was never to the taste of Cardiff. Worthington 'Dash', as it was known, proved hugely popular in Swansea, but in the capital Brain's Red Dragon now dominated the dark market.

Even the brewery's speciality of light mild was no longer safe, with draught pale mild being shipped in from Bass's Springfield Brewery in Wolverhampton for racking in Cardiff. Tank PMA came from Burton. But Cardiff reasserted itself in this area, with a bewildering array of brands on offer, such as Hancock's PA, Welsh Brewers' PMA, Worthington PA and Worthington M. But it was all a liquid illusion at the bar. As Bass admitted in 1978, they were all essentially the same brew. Cardiff also took over brewing of Worthington Best Bitter. But Allbright was now the star brew, living up to its reputation as 'the most popular pint in Wales'. In the 1990s Bass claimed it accounted for one in every five pints of ale drunk in South Wales. Sales were helped by the fact it was now branded a bitter.

High-gravity brewing was adopted, which meant the beers were produced in concentrated form and then watered down to the appropriate strength. This allowed the Cardiff site to boost production to approaching a million barrels a year. It was a far cry from William Hancock's first years, but a reminder of the past briefly rose into view in 1996, when his early phoenix symbol was used on the pump-clip of a special dry-hopped version of HB called IPA. It was a short-lived flight, once Bass decided to sell its Cardiff brewery. Now only HB is left to fly the flag for Hancock's at the bar – brewed by Brain's at the former Hancock's brewery.

you can't
start
better

than with a 'bitter'

ELy

SPECIAL PALE ALE
ON DRAUGHT
1/5d. a pint   ON SALE AT ALL OUR HOUSES

Above left: Only HB still carries Hancock's name today.

Above right: Bitter was often sold at a higher price to mild, even when it was weaker, as in this 1958 advert.

Ely Beer Of Good Cheer

When the Ely Brewery Company began brewing in 1888, its first pints were just what local drinkers wanted. There were two milds, XX with a gravity of 1046 and XXX at 1054. Though Bitter Ale (1048) was weaker than XXX, it sold at a much higher price, costing publicans 38s a barrel, compared to 32s for XXX and 28s for XX. Bitter was regarded as a premium product compared to mild. Little changed over the following decades apart from the strength of the beer and the closing of the price gap between XXX and BA, with new premium beers introduced.

When Rhondda Valley Breweries took over in 1920, the directors instructed the brewers at its three plants in Ely, Pontypridd and Treherbert to brew the following beers: XX at 1033 gravity; XXX and Bitter Ale (both 1040); and Pale Ale and Stout (both 1046). XX was to sell at 6d a pint, XXX and Bitter Ale at 7d and Pale Ale and Stout at 8d. Scribbled in the margin of the directors' minutes is the note: 'decision of South Wales Brewers' Association, April 1920'. The company was not only co-ordinating its own production and prices, brewers across the region were also falling in line with each other. The industry always strongly denied that it acted as a cartel, but price rises were regularly synchronised. This was why the established brewers resented the arrival of clubs breweries after the First World War, which were outside their control.

During the conflict, many clubs had not been impressed by the weaker beers, much higher prices and poor service, with brewers looking after their own pubs, rather than clubs, when

supplies were short. The South Wales Club and Institute Union decided to do something about it. At a meeting at the Cathays Liberal Club in Cardiff on 28 March 1919, members resolved to set up their own brewery. In June they bought the Crown Brewery at Pontyclun and established the South Wales and Monmouthshire United Clubs Brewery to supply clubs. The clubs ran the brewery as a co-operative, so prices to members were low. Its dark mild, XXXX, sold at just 3d a pint, half the price of Ely's cheapest beer, XX. Ely, which had most of its pubs up the Valleys where the clubs were concentrated, was hit hard. By the end of the year Ely had stopped brewing XX. It was not an auspicious start to the troubled inter-war years and, after a brief boom in the early 1920s, the company struggled to survive.

Then came Lazarus Nidditch after the Second World War, who not only revived the firm, but transformed its ales. The growth market in the 1950s was in bottled beer and Nidditch rattled the beer crate with a cascade of new brews, with fresh titles, innovative labels and catchy slogans. It was a revolutionary approach. At the time, most breweries produced a standard range of beers with standard names like brown ale, pale ale, strong ale and stout. Usually the company had just one set

Opposite: Lazarus Nidditch was determined that Ely's bottled beers would stand out, with colourful labels, individual names and powerful adverts like this one for Strong Ale in 1954. Golden Gleam and Druid's Ale were two of the new distinctive bottled brands.

Right: Brewer's Own was marketed across the UK, with the bold slogan 'None Better in Britain'. Pottery figures were even produced to decorate bars, showing a brewer admiring a glass.

design for its labels, with the only difference between each brew being the colour of the label and the generic name of the beer. In the 1950s both Brain's and Hancock's fell into this comfortable category. Brain's oval labels all featured their familiar red dragon, while Hancock's carried John Bull. They were easy to recognise as belonging to the brewery, but hardly inspiring. Nidditch was not impressed. He wanted something with much more impact. He wanted his bottled beers to stand out from the crowd at the bar. And each one to stand out from the others.

The idea of individual beer brands was not new. In the early 1890s, the County Brewery in Cardiff had tried to develop sales of its light bitter under the name Golden Hop. But it does not seem to have been a success, and the brewery was soon taken over by Hancock's. In a highly conservative industry, the main brands were brewers' names, notably Bass and Guinness, which came to represent particular beers. But after the Second World War, as competition increased, more enterprising brewers were looking for something to give them an extra edge. Individual beer brands were one answer and Lazarus Nidditch grasped the idea with enthusiasm. He was well ahead of most of his rivals in spinning the bottled beer trade.

Landlord Mr King pulls a pint at the Junction Hotel, Taffs Well, just outside Cardiff, in front of his grand Christmas display in 1957. And he only came fourth!

In 1948 alone three new bottled beers appeared – Golden Gleam, Golden Harvest and his great bright hope, Brewer's Own, which he aimed to develop into a major brand across the UK. This premium pale ale was marketed under bubbly, boastful slogans like, 'The Champagne of Bottled Beers' and 'Best in Wales – None Better in Britain'. Its distinctive label showed a traditional red-capped brewer gazing in admiration at a glass. It was a premium beer at a premium price. In the early 1950s it sold at 1s 4d a half-pint. The next most expensive bottled beer, Imperial Pale Ale, cost 10d. The name, Brewer's Own, was protected as a trademark and it was advertised widely across the UK, even in the *Financial Times*.

As the brewery boasted in a leaflet produced in 1952, showing off the range of new labels: 'The popularity of Ely bottled beers is growing throughout South Wales, and more people are daily asking for various Ely brews. One of the contributing factors towards establishing the popularity of these beers and making for expanding sales are the brand titles and the distinctive publicity given to them.' Bottled beer sales surged from a tiny 0.5 per cent of production in 1947 to 15 per cent by 1957. A new bottling stores had to be built the following year to keep up with demand.

But not all the new bottled brews were a hit. Volatile Lazarus Nidditch kept changing his mind. Golden Harvest and Golden Gleam failed to shine and were soon dropped. Druid's Ale tried to evoke Wales's past but was eventually replaced by potent Little Gem, 'a special winter ale'. Both were sold in nip-size bottles. At the other end of the strength band, a light brew, TV Ale, was successfully launched in 1953 in two-pint flagons to meet the growing challenge of pub drinkers staying at home to watch the box. The same year, Ely was the only Cardiff brewer to produce a special bottled Coronation Ale. In 1954 Strong Ale flexed its muscles with a weightlifter on the label. Established bottled brews like Home Brewed were also relaunched with new labels and slogans. In 1956 a glucose stout was introduced, but was swiftly replaced by stronger Gold and Silver Stout the following year. Silver foil was increasingly used on the bottles and landlords were encouraged by the brewery to put on colourful bar shelf displays to boost sales, with cash prizes for the best efforts.

During the frenzied focus on bottled beers, Ely's draught beers at the bar had barely changed. Traditional XXX Dark and BA Bitter were still the mainstay of the brewery, with Special Pale Ale a couple of pence more per pint. There was little to excite Lazarus Nidditch – until the arrival of filtered and pasteurised container beer. A few brewers, notably Flowers of Stratford-upon-Avon and Watney's of London, had been experimenting with 'keg' beer since the 1930s, but the dispense equipment had kept letting them down. Then in 1958 a new design of keg allowed sales to froth up, led by Watney's Red Barrel. Lazarus Nidditch jumped on the dray wagon at once.

The same year Ely launched Silver Drum with the slogan 'You can't beat it'. It was sold to publicans and club stewards as a consistent, convenience beer that could be served at once without lengthy conditioning in the cellar. The nine-gallon metal kegs and accompanying carbon dioxide cylinders could be fitted in bars which could not handle traditional cask beer. It was sold to drinkers as a premium product, served through a large silver font on the bar. Adverts proclaimed: 'Something new in draught beer – a sparkling pale ale matured to exactly that state of prime condition demanded by the beer connoisseur.'

In fact it was just a fizzed-up and filtered version of the SPA – sold at 3d to 5d more a pint. But drinkers were sold on the new presentation. Even the beer mats were different, being stamped in silver on coloured lino. Silver Drum was sending out new marketing messages, and when Ely Brewery was snapped up the following year by Rhymney, it was one of only two beers to survive the takeover. The other was the premium bottled beer, Brewer's Own.

Rhymney & Crosswells – The Best Round Here

When Crosswells was formed in Cardiff in 1892, the company was noted for being bottlers of Guinness. And even after it started brewing in Ely in 1901, its reputation always shone strongest in glass. After the takeover by Rhymney in 1936, the men from the Valleys at various times considered closing the Cardiff brewery and just concentrating on bottling at the site. Its future was always closely tied to its bottling lines and the beers that flowed from them.

Crosswells marketed thirteen different beers in the late 1930s. But all was not what it seemed. There were only essentially three different brews, as one of its former workers later disclosed. Bob Pritchard had joined Crosswells as a pupil brewer in Cardiff in 1935, becoming second brewer by 1939. He rejoined the company after the war, before leaving to join Edme, which made home-brew kits. When he wrote a book for Amateur Winemaker in 1983, called *All About Beer*, he revealed some tricks of the trade.

He said Crosswells brewed three pale beers in Cardiff, an IPA at 1048 gravity, a Bitter at 1041 and a Light Bitter at 1035:

> From these three basic brews, no less than thireen different types of beer were sold from the brewery. This was done by the addition of caramel [burnt sugar]. This was added, in varying degrees, to subdivisions of the three brews. In draught form, the three went out as IPA, Bitter and Light Bitter and, coloured, as Strong Ale, Homebrew and Mild. This was repeated in the bottling department and six bottled versions were produced.
>
> The thirteenth type was Oatmeal Stout in bottle and, for this, heavy colouring was added to the 1048 IPA. One pound of oatmeal was always added to the IPA mash, which made no difference at all in the brew of 200 barrels, but Excise insisted on it to enable us to use the word oatmeal on the label. They were all excellent beers and the Oatmeal Stout was extremely

Whitbread were alarmed when they discovered how Crosswells' Oatmeal Stout was produced.

Chestnut Ale was appreciated by drinkers, despite Whitbread's reservations about Crosswells beers.

Kegs of Hobby Horse are rolled on to a dray at Crosswells Brewery on a freezing February day in 1963.

popular. The 1048 IPA was a good, full-drinking, hoppy bitter and the large addition of caramel, with its acridity, cut across the original flavour and made a really good stout.

The brewery was eventually taken over by a London brewery [Whitbread in 1966] with a quite well-known stout of their own, and they threw their hands up in horror at the 'IPA converted' product; it was withdrawn and their own was substituted. We were then inundated with letters of protest demanding that Oatmeal Stout be brought back!

Another popular bottled beer was Chestnut Ale, while from Rhymney came a powerful golden barley wine, King's Ale, which was first brewed for Edward VII's Coronation in 1902. But the Hobby Horse brewers were also prepared to canter into unknown territory. Besides producing their own stronger keg beer, Hobby Horse, in 1959, they also ventured into areas few brewers at the time had considered – such as producing a bottled beer specifically aimed at women. Tivoli was introduced in the late 1950s with the slogan 'The modern ale for modern women', with a fountain design on the label. Just to hedge their bets, Rhymney added the rider 'Men like it too'. After taking over Ely, the group also slalomed into the fledgling lager market with Ski Lager in the early 1960s, with the yeast flown over from Pripps Brewery in Sweden. Lager was one of new chairman Harry Llewellyn's favourite drinks, but the rest of the bottled beer range was reduced down to Welsh Brown, Welsh Bitter and Welsh Strong.

Wooden casks are filled at the brewery in 1963, but Crosswells was mainly known for its bottled beeer.

When Whitbread eventually took total control in 1966, most of their bottled beers were already well established at Crosswells Brewery. Their draught beers, Trophy and Tankard, now rolled in to replace most of the Rhymney brews. When Whitbread called time on its Cardiff brewery in 1982, production of the last brew with a local connection, keg Welsh Bitter, was transferred to the giant Magor Brewery. The closure was marked with a heavy drop of the last Whitbread Gold Label barley wine to be bottled there.

four

THE PUBS

When Ben Davis, a leading architect for Allied Breweries, wrote *The Traditional English Pub: A Way of Drinking*, published in 1981, he opened the book by saying, 'The pub is English. The Scots drink in bars, and the Welsh, bless them, will drink anywhere.' In a sense he was right. For pubs were everywhere in the centre of Cardiff and the docks.

Wakeford's Cardiff Directory of 1863 lists 211 hotels, inns and public houses. Forty years later the number had risen to 274, with the vast majority crammed into the central and Butetown areas, where 180 licensed premises served a resident population of less than 25,000. In sharp contrast, the suburb of Roath, with a slightly larger population, had just twenty-six pubs. Drinkers in Cathays struggled to find a drink at all. Almost 40,000 people were served by a mere nine houses. Canton was better off, with forty-two pubs for nearly 37,000 residents.

The reason we know so much about Cardiff's pubs in 1903 is that in January the police were sent on a pub crawl. They were not looking for illegal gambling or drinking on Sunday. Instead they went armed with a check list and a tape measure. Chief Constable William McKenzie had told the Cardiff Watch Committee in December 1902 that 'the closing of unnecessary public houses in certain congested districts would materially assist in reducing crime'. Because there were too many pubs, he believed some landlords could not make an honest living. 'Some licensees have told me that if they were to exclude loose women, they could not pay their way.'

There was nothing novel about this concern. Back in 1790 the Grand Jury of the County of Glamorgan was worried that 'an excess in the number of alehouses is a grievance which should be redressed, but whether twenty-seven are too many to answer the convenience of the town of Cardiff and its neighbourhood, we find ourselves incompetent to determine from want of sufficient evidence on that head.' More than a century later, the licensing magistrates were still uncertain. So they asked the police to investigate the extent of pub overcrowding, and the facilities each house provided. The result was that on 14 January 1903, policemen across Cardiff set off on their pub crawls, to pace out the distance from bar to bar and gather information. This was no leisurely stroll. Within ten days they had compiled a detailed 277-page report.

The sign on Sam's Bar at the end of St Mary Street recalls its time as the Terminus, part of Cardiff's four-bar trick (see page 114).

It demonstrates just how many licensed houses were squeezed into some areas. If drinkers at the Bridgwater Arms in Bute Terrace were not happy with the beer, they did not have far to walk to the next inn. The Merthyr and Dowlais was 39 yards away, the Prince of Wales 100 yards, the Wheatsheaf 114 yards, the Windsor Arms 122 yards and the Tredegar Arms 150 yards. Many streets were tight with havens for loose women. Adam Street boasted seven pubs; Bridge Street had eight, while Bute Street, the long dive to the docks, offered a staggering thirty-one. Some were crammed into very tiny pockets. Drinkers at the gaudy Golden Cross at the top of Bute Street had to lurch just one-and-a-half yards to the next pub. The Fishguard Arms was right next door. While the White Swan was 14 yards away, the King's Arms 24 yards, the Charing Cross 32 yards and the the Salutation 35 yards.

The report also reveals what each pub offered the public – to the exact inch. The Bridgwater Arms had a 35ft 1in by 10ft 5in bar with two partitions, a 20ft 8in by 11ft 3in smoke room and a 13ft 5in by 19ft 4in club room. There was also a WC and a urinal. And that was just on the ground floor. On the first floor was a larger 41ft 4in by 13ft 8in club room. Temperance campaigners would have grimly noted that the pub offered no refreshments other than intoxicants, and that it was tied for its beer supplies (to Brain's). With keen competition on the doorstep, it was clearly not an easy business. The licence had changed hands six times in the previous eleven years.

And the 274 pubs listed were just the ones with licences. Many private houses opened their doors on a thirst come, thirst served basis on the dry Sabbath. Illegal drinking dens or shebeens were widespread, and many official clubs were little better. If Cardiff city centre has a riotous reputation for heavy drinking today, it was no different a century ago. When the Bridgwater Arms, built in 1881, closed in 1948 and the canal opposite was filled in to make way for Churchill Way, workers excavating the waterway made an odd discovery. In the mud under the low canal wall near the pub door, they found dozens of pairs of false teeth. Many drinkers, worse for wear after a good night out, had lost more than their lunch when they leaned over to admire the dark depths below.

Two hundred years ago, when Cardiff was little more than a posting station on the route to West Wales, the coaching inns knew that travellers wanted to soothe the bumps of the rutted road with a stiff drink or two. When Thomas Thomas took over the Rummer Tavern and Steam Packet Hotel in Duke Street, pride of place on his poster went to declaring that he offered 'A choice assortment of excellent wines, prime spirits and home-brewed beer'. Then Cardiff's hostelries were split between the grander inns by the castle, such as the Cardiff Arms Hotel, which gave its name to the famous rugby ground, and the more humble houses serving the small market town. Few of these still stand today. Even the imposing Cardiff Arms was demolished in 1882 for road widening. And where the names survive, like the Rummer Tavern, the pub has been completely rebuilt.

For Cardiff was substantially a Victorian creation, and the town expanded so rapidly that often pubs were rebuilt more than once to keep pace with the booming population. Sometimes the only way was up. The narrow Borough Arms in St Mary Street shot up from a three-storey

The Royal Oak in St Mary Street in the 1880s, a dilapidated remnant of Cardiff's past, with the Blue Anchor on the right.

building to double the height in the late 1880s, having been dwarfed by the surrounding market buildings and Howell's department store. Its outside design shows the liquor industry's growing confidence, with a bulging bow window on the first floor and fancy pillars and flourishes on the towering structure.

Nowhere is the crowded pub culture of Cardiff better illustrated than at the bottom end of the same street. For there once stood four pubs right next door to each other, jostling together for business like four customers eager to get served at the same bar. The four were the Royal Oak, the Blue Anchor, Elliott's Hotel and, on the corner, the Terminus. All four buildings still stand today, though only one, the Terminus, remains as a pub. Their development mirrors the changing nature of drinking in Cardiff, and the changing face of the city.

An early photograph from the 1880s (see previous page) shows the Royal Oak as a dilapidated remnant from Cardiff's past, a crumbling reminder of the days when Cardiff was a modest market town. Its founding as an inn predated the licensing register. The pubs at the bottom of St Mary Street were clustered just inside the South Gate, which guarded the road to the sea. Already the Blue Anchor, on the right, first licensed in 1849, has grown to much grander heights. But despite its rundown appearance, the Royal Oak classified itself an inn rather than a public house. Landlord Thomas Price offered accommodation and food like cold meats and bread and cheese. The Royal Oak was determined to rise to the challenge of its next-door neighbour.

The photograph below shows the three-storey brick structure which replaced the old Oak, with elaborate window and door arches and gargoyles under the eaves. This was the era when

The Royal Oak after rebuilding in the 1890s.

many pubs were still free houses, before the local breweries snapped them up. They often offered beers from distant breweries. Mitchells & Butlers ales came from Birmingham. The food had also improved. The menu on the board by the door was kidney soup for 2d, roast beef and two veg for 6d and apple pie for 2d. The Blue Anchor next door also served hot dinners, soups, Bovril, tea, coffee and sandwiches. It is a fallacy that pubs have only started providing meals in recent decades. In 1903, only 26 out of 274 houses in Cardiff offered no food at all. They were the workers' restaurants as well as their watering holes.

At the other side of the Blue Anchor, Elliott's Family Hotel, first licensed in 1872, towered above its neighbours, five storeys high. It prided itself on being a proper hotel, providing twelve guest rooms with thirty beds, with family sitting rooms and coffee rooms, besides having a bar and smoking room on the ground floor. Across the road it had substantial opposition in the striking Great Western Hotel on the approach to the railway station and Barry's Hotel, which was famous for its steak and kidney puddings.

Then came the Terminus, the only surviving pub, which owed its name to the horse-drawn buses which left here for Penarth. Like the Royal Oak, it was pouring pints before records began. In 1903, there was a warren of walls on the ground floor, with a public bar divided into three separate compartments, plus a lunch bar, a smoking room and a billiards room. Given its prominent corner site, it was an ideal position for a new brewing company to announce its arrival. In 1888, the Ely Brewery Company bought the Terminus for £2,000. By 1903, the next door Elliott's Hotel was also a tied house – to Ely's close competitor Crosswells. The hotel at

The Royal Oak, the Blue Anchor, Elliott's Hotel and the Terminus in operation next door to each other at the end of St Mary Street, on the right, early in the twentieth century.

No. 62 St Mary Street became the brewery's central Cardiff offices. This prompted Ely to plaster the Terminus's walls with slogans proclaiming 'The Original Ely Beers and Stout'. The bitter rivalry in Ely had been transferred to the bottom of St Mary Street.

Surprisingly, none of the four pubs fell into the hands of Cardiff's most dominant brewers, Brain's and Hancock's. The Royal Oak and Blue Anchor later also joined the Rhymney & Crosswells empire. In the 1950s both sported Rhymney's hobby horse sign in competition with the Ely barrel on the Terminus, Elliott's having been turned over entirely to offices. Then the galloping hobby horse captured Ely, and Rhymney ruled the block, owning all three remaining pubs. After another takeover the pubs passed to Whitbread in 1966, but the London brewing giant had no interest in investing in three houses so close to each other. Within a few years Whitbread had disposed of the Royal Oak and the Blue Anchor, which both became restaurants. Only the Terminus, later known as Sam's, remains to remind drinkers of Cardiff's remarkable four-bar trick.

Just a short stroll round the corner is a much more stylish survivor, though it needed a determined campaign in the late 1970s to prevent it from being demolished. The Golden Cross is Cardiff's most palatial pub, with a colourful tiled façade. But its real beauty lies within, with its elaborate green and gold glazed bar and tiled murals of Cardiff Castle and the old Town Hall. It's an ornate echo of a brash, more confident era. It stands proud at the top of Bute Street, gazing down into the docks, and its equally colourful history links the city centre with its murky waterfront.

The present Golden Cross is an Edwardian creation, having been rebuilt in 1904, though there has been a pub on this site since 1849, originally known as the Shield and Newcastle Tavern and then the Castle Inn, before becoming the Golden Cross in 1863. Despite its religious name,

The three surviving pubs in 1968, now all belonging to Whitbread. Soon two would become restaurants

Even the grand Golden Cross used to have another pub in its back pocket, the Fishguard Arms.

it was a far from holy place. Its rebuilding only moved it upmarket. It offered a lively night on the tiles for sailors off the ships and developed a reputation as the smartest brothel in town. Its alluring atmosphere was laced with legends. One of its red-light regulars, Carrie Gilmore, was notoriously stabbed to death in West Canal Wharf in 1907, and her murderer was dubbed the Cardiff Ripper. Oswald Mosley tried to hold a meeting of his black-shirted fascists in the pub before the Second World War, but his car was overturned by local protesters, and he retreated, bruised and battered, back to London. During the war American servicemen enjoyed the delights of the Golden Cross and one young GI is alleged to have swung his first fist in anger during a fight in the bar. His name was Rocky Marciano.

In 1978 it seemed this flawed gem was about to disappear when South Glamorgan County Council proposed to drive a road through the site. But a vigorous campaign led by the *South Wales Echo*, the Cardiff branch of CAMRA and Brain's saved the listed building. The brewer promised to spend £100,000 restoring the pub to its former glory. It later must have wished it had not fought so hard to save the landmark, for the final cost was four times that. When work began, it was discovered that the glossy boozer was unsteady on its feet. It had been built without proper foundations, so first the whole building had to be underpinned, and the walls at the back rebuilt. The weight of a cast-iron bay window on the first floor was found to be damaging the structure and so it was replaced by windows matching the rest of the building. The bay window had been above a betting office, which was now incorporated into the pub, while the neighbouring Fishguard Arms was demolished to provide a small delivery yard. Only then could painstaking work begin on the tiles, with specialist companies brought in to replace damaged panels. The original murals, made in Jackfield, Shropshire were restored, and a new bar added at one end which had formerly been a counter at Cardiff's old central post office in Westgate Street. A cast-iron fireplace, which had been upstairs in the living quarters, was brought down into the bar. Two gas lamps at the new side entrance were rescued from the demolished Barry Dock Hotel.

The magnificent tiled bar and wall murals inside the Golden Cross, after restoration in 1986. Landlord John Gallagher and his fiancée Christine Williams planned to have waitresses in Victorian costumes and offer cocktails, but the idea did not last long.

Tom Jones and the Treorchy Male Voice Choir turned up to test the acoustics and Brain's Dark in 1987. Both went down well.

'The refurbishment of the Golden Cross is the most complicated and complex operation of its kind that the company has ever undertaken,' said Brain's estates manager Ian Richards when the pub finally reopened in 1986. The restoration won the Prince of Wales Award (for service to the community) the following year, the first time a pub had received this honour.

What is just as remarkable is that the Golden Cross seems to be the only shining example in Cardiff of a 'gin palace' (its exterior tiles significantly advertise wines and spirits rather than beer). Other cities which sprang up in the Victorian era like Liverpool and Birmingham have many more 'gaudy, gold-beplastered temples', their interiors full of carved wood and engraved mirrors. But the powerful temperance movement and disapproving licensing authorities frowned on such extravagance in Cardiff. 'Welsh pubs suffer under the heavy hand of Puritanism; they are tolerated rather than accepted,' commented architectural historian Brian Spiller in 1952.

Though some houses, particularly those on the main roads out of the city, have imposing exteriors, they were usually more modest inside. Cowbridge Road, heading west out of the city, is a classic alcoholic artery, fronted by pushy pubs, many erected to impress in the late nineteenth century. From the elegant Wyndham Hotel built in 1881, past the Royal Exchange (1887), with its pillared, arched windows, to the extravagant Corporation Hotel (1894), each seems to grow grander as the century advances. Even the bold, brick Duke of Wellington, built in 1892 on one corner of Brain's Old Brewery site, was described in 1983 by one regular of fifty years, as 'a real spit and sawdust pub'. The brewers probably felt they had no need of extensive brasswork, etched glass and rich rosewood to ensure locals drank their beers.

Pubs on busy roads tried to stand out from the surrounding houses, like the Plymouth in Grangetown, pictured in 1968. But most had more modest interiors.

In the docks a few pubs, like the long-demolished North and South Wales in Louisa Street or the surviving Packet in Bute Street, had finely crafted backpieces behind the bar. The uncompromising Mount Stuart opposite the dock gates had no fancy flourishes on the outside, but this solid block of a building concealed mirrors inside reflecting the prosperity of the area when coal was king. In the back room, ship owners, sea captains and shipping agents sealed their deals alongside an ornamental bar screen. While the front bar heaved with dusty dockers, coal trimmers and sailors, as tight as a barrel of herrings. When this bare bar shut for the afternoon, it was claimed the barmaids could sweep up enough coal to keep the fires burning. You knew the old docks had finally died when the Mount Stuart was closed in 1985.

Divisions were sharp and salty on the seafront and the bars rarely mixed. The famous Big Windsor is a classic example. After the war it developed a reputation for its gastronomic delights when French chef Abel Magneron leased the waterfront pub. Celebrities like Noel Coward and Katherine Hepburn dropped in to try the menu. Yet a small side door led to another world – the 'Snake Pit'. There, among the sawdust and spittoons, a very different crowd enjoyed the rough cider. Around the grand Coal Exchange, where fortunes were made and lost in an afternoon, seamen of all nationalities sprawled from the bars. While inside the exclusive Exchange Club successful dealers played skittles with champagne bottles at three o'clock in the morning.

A few pubs in the city centre also hint at extravagance. The Oxford Hotel on The Hayes, which advertised itself as a wine and spirits vaults in the 1890s, presented a fine façade at ground level with marble columns when first built. It was demolished in the early 1960s to make way for Oxford House and Arcade. The tiles on the Queen's Vaults in Westgate Street are older than the Golden Cross, but its interior has been gutted so many times, it is impossible to tell how ornate it was originally inside.

But there was one other prominent pub which more than matched the Golden Cross, though its glories are now hidden from view behind a burger bar. And like the Golden Cross its foundations were soaked in wines and spirits as well as beer. The Green Dragon, on the corner of Duke Street and St John Street, was one of Cardiff's oldest pubs. In 1859 Scotsmen Andrew Fulton and Matthew Dunlop started to trade as wine, spirit, ale

The Big Windsor, seen here in 1988 looking over to Penarth, fell into disrepair in the 1990s and eventually became another waterfront restaurant.

and porter merchants in Duke Street and took over the Green Dragon as their showcase retail outlet. Known simply as Fulton Dunlop, they decorated the premises in grand style to ape the extravagant interior of nearby Cardiff Castle, crafted by eccentric architect William Burges from 1865. Burges' love of medieval mysteries and exotic animals is particularly reflected in the Mahogany Room, developed in 1905, where drinkers could enjoy glasses of Draught Bass beneath carved wooden columns, stained-glass knights and galleons, mosaics of birds and fish, and a high, arched ceiling. The room was so stunningly superior, councillors often preferred it for informal committee meetings to City Hall. Sadly, the corner premises were taken over by the neighbouring gas showrooms in the 1960s and in 1988 became the largest Wimpy outlet in the UK. Today it belongs to Burger King. But no one could bring themselves to remove the Mahogany Room, and it remains locked away, Cardiff's hidden secret, an oasis from another age above the frantic fast-food trade below.

It was the last pub to go on what had been another crossroads of cheer in the city, where the road north to the coalfields crossed the main east-west thoroughfare, which had once boasted a pub on three of its four corners. Directly opposite Fulton Dunlop's had been the Law Courts Hotel, originally the Glove and Shears, which used to carry a huge advert for Worthington along the length of its side wall. It was demolished in 1923, with the rest of the northern side of Duke Street, for road widening. On the other side of North Street (later Kingsway) stood the Red Lion, which was taken over by the neighbouring Roberts store after the Second World War, before also being reduced to rubble. And just a few doors up from the Red Lion was the Rose and Crown, which had greeted visitors from the Valleys to the town since at least 1787. With its cobbled courtyards and extensive stabling, including a gymnasium, it was one of the landmarks of the city. It was demolished in 1974 for more road widening, and replaced by a subterranean bar, later known as Coopers and now the Barfly. The motor car has a lot to answer for, as it has driven much of the character out of Cardiff.

Two pubs did survive close to Fulton Dunlop's. Almost next door in Duke Street was the distinctive half-timbered Rummer Tavern, which at times in its long history was known as

Above: An early Fulton, Dunlop label.

Right: The Mahogany Room, now hidden away above Burger King.

Below: Kingsway was already busy with cars just after the Second World War. The Red Lion, left, would eventually be demolished for road widening in 1974. Fulton Dunlop is on the corner, with the half-timbered Rummer Tavern (Hallinan's) almost next door.

Hallinan's after the family who owned it for many years. Like its near neighbour, it also once boasted an upmarket, upstairs bar known as the Oak Room because of its fine wooden panelling. While round the corner in St John Street was the much larger, sweeping front of the Tennis Court Hotel, originally the Kemeys-Tynte Arms, later the Buccaneer and now the Owain Glyndwr, though it has changed its name and interior so many times, it is doubtful whether anything original survives apart from the outside walls.

Many of Cardiff's pubs had been under threat since the start of the twentieth century, when the chief constable had called for fewer licences in congested areas. Heavy drinking was certainly a major problem. In the first six months of 1852, 80 per cent of proceedings at Cardiff Borough Police Court were for drunkenness. And the problem was getting worse. In 1884, there were 958 cases in Cardiff. By 1897 the figure had hit a peak of 1,667. Around a quarter of the offenders were women. It is another fallacy that women never used pubs, except as barmaids or 'ladies of the night'. They just went into different bars within a pub to men, meeting in the snug or sending children for carry-outs from the jug-and-bottle hatch. In 1918, the South Wales brewers were forced to send a letter to their tenants 'to prevent women congregating in licensed houses', telling them to tackle 'increased drinking among women and, if possible, put an end to this evil'. They were instructed to serve women only between 6 p.m. and 9 p.m.

Temperance campaigners vigorously backed the long arm of the law in shutting bar doors for good. Preacher Eben Fardd branded the increasingly imposing pubs as 'highly mischievous, rearing their aristocratical fronts in the midst of a humble, industrious and poor population and irradiating forth their devilish attractions to sap the foundations of social virtue.' They tended to ignore the fact that these temples of temptation offered the only escape for many from a hard

Women certainly felt at ease in this Adamsdown local in 1961.

life. Few workers had much in the way of home comforts, and often the pub was the only local social centre, providing companionship and entertainment as well as food and drink.

Though the First World War sharply reduced both the strength of beer and pub opening hours, the police at first remained hostile to the licensed trade. This was the era of regular raids on bars and laughable court cases. Most concerned drinking after hours. 'I peered through the bar window at 11.13 p.m. and saw two full glasses on the table, your honour,' straight-faced policemen would tell the bench. Then there was gambling. On 24 November 1923, Inspector Preece entered the Roath Castle in Castle Road, Cardiff, after hearing the shout 'Roll up for the rabbits'. In the bagatelle room he 'found a number of men rolling balls along the table, the scores being assessed according to the pockets they entered'. From the mantelpiece were hanging the winner's prize – two dead rabbits. Landlord Samuel Andrews' defence was that gambling was 'going on everywhere'. Though he admitted allowing gaming, magistrates dismissed the prosecution. They were tiring of these cases taking up court time. The same year a plain-clothes policeman had been sent undercover to a whist drive at the Cardiff Arms Hotel in Treorchy. Magistrates amused themselves by adding up PC Miles' score card and telling him he could have won a prize. Landlord Thomas Williams was given a nominal fine, while the *Western Mail* called for a review of the licensing laws. 'There's no reason why the law of the land should cut across the rules of common sense.'

But the police were under pressure from the moral middle class. When Chief Constable James Wilson proudly reported to the annual licensing session in February 1925 that cases of drunkenness the previous year had fallen by 49 to 371, he was facing the massed ranks of the Cardiff Citizens Union, the United Crusade of Religion and Morals for Wales, the Free Church Council, the British Women's Temperance Association, 'various sisterhoods of the city' and many ministers. That year he played to the gallery and asked for a further ten Cardiff pubs to have their licences withdrawn, including five alone on Adam Street – the Rhymney, Eagle, Vulcan, Sandon and Wheatsheaf. He also wanted the Crichton Arms and East Dock in Tyndall Street declared redundant, along with the Royal George in Roland Street, the Crown and Sceptre in Pendoylan Street and the Duke of Edinburgh in Ellen Street. Then the breweries would have to desperately barter to save the licences of the pubs they most valued. The previous year the Ely Brewery had managed to save the King's Castle on Cowbridge Road, which did 'a substantial trade', by promising structural improvements.

Even then the moral majority was not satisfied. The Free Church Council objected to the licence for the Angel Hotel in Westgate Street, one of the most prominent hotels in the city. Actions like this, now that the city's drunks had mostly dried out, meant that support for the hard-line temperance movement began to drain away. Many leading residents were not impressed when teetotal Alderman C.F. Sanders became Lord Mayor in 1932 and promptly pledged it would be 'a dry civic year' with no alcohol at receptions during his term of office, including the Lord Mayor's Banquet at the City Hall. And campaigners invited ridicule when the council's tramways committee the same year tried to stop pub advertising on the back of bus tickets (beer advertising was already banned). 'We're rather tired of the continued attempts to use the council chamber as the medium for furthering teetotal propaganda, much of which in the past has made Cardiff the laughing stock of the country,' complained Cardiff Licensed Victuallers' Association. The full council overruled the decision. The tide was turning.

Chief Constable James Wilson demonstrated the huge change in attitude when he praised Cardiff licensees in February 1933, saying their pubs were 'well conducted' and that they had obeyed the law despite poor trade. The *South Wales Echo* was so stunned it made his report to the licensing justices the front-page lead. Drunkenness cases had dropped again by 56 in 1932 to 158, with 101 different men convicted and 30 different women. The number of pubs in the

Chief Constable James Wilson caught everyone by surprise by praising Cardiff's publicans in 1933, as *South Wales Echo* cartoonist J.C. Walker illustrates.

city had fallen by almost a quarter in thirty years, down 69 since 1903 to around 200. Instead the police chief focused his fire on the clubs, which he said had become the new shebeens. The number of licensed clubs in Cardiff had almost doubled in ten years, from 31 in 1922 to 60 in 1932, with almost 18,000 members. The attraction of clubs was that they could open on Sunday. Some were 'out of control'. The Victoria Park Social Club, only formed in 1929, was struck off the register in 1932 because of 'frequent drunkenness' and 'illegal sales'.

In 1934 the chief constable even spoke up for women drinkers: 'There is no reason to believe that the patronage of public houses by females is likely to decrease. In fact, it is increasing and likely to continue to do so. I therefore respectfully suggest that the time has arrived for the committee to consider and, if they think fit, to promulgate a policy regarding proper amenities for women at licensed premises.' It looked as if the days of dodging into the Gents were over. Women were about to get their own toilets.

The brewers welcomed the suggestion. Shaken by the threat of prohibition in the First World War, they had been working hard to improve their reputation and their pubs. Brain's employed leading architect Sir Percy Thomas to rebuild some of its pubs, such as the Westgate Hotel and the Birchgrove, in grander but restrained style. Gone were the wall paintings proclaiming 'It's BrAIn's You Want'. In their place were buildings which looked like street-corner country mansions, with better facilities and novelties like comfortable seating.

The pub was probably in its prime between the wars. With television yet to switch many on to staying at home at night, it was the heart of the local community. It did not just provide popular bar sports like darts, dominoes, draughts, billiards and cards, plus the Cardiff speciality of skittles – many pubs had an alley – it was also the headquarters of everything from local rugby, football, and cricket teams to cycling, drama and rambling clubs. The brewers encouraged local sports leagues to boost interest. The Boar's Head in Canton in the 1920s was typical. Besides hosting

The Westgate was demolished in the early 1930s.

Its replacement, pictured here in 1957, was widely praised for its restrained design by Sir Percy Thomas.

many rugby teams, it was also the headquarters of the Canton baseball, air rifle, pigeon and bagatelle clubs. While a huge variety of charities were supported, not just with boxes on the bar, but organised fund-raising events. Regulars were helped with loan clubs and Christmas clubs.

From being officially regarded as the enemy behind the lines in the First World War, where soldiers and workers might be led astray, the pub became the blockhouse on the home front during the Second World War, where morale was maintained. 'In these days, it seems to me, the British pub, the people's club, has justified its existence as perhaps it never did before,' wrote the MP A.P. Herbert during the conflict. 'For it has been the one human corner, a centre not of beer but bonhomie; the one place where after dark the collective heart of the nation could be seen and felt, beating resolute and strong.' Pubs like the Neptune in Caroline Street had been requisitioned as recruiting centres in the First World War. In the Second World War, virtually every bar eagerly set up savings stamps groups to support events like Warships Week or the 'Tank'ard Fund. Posters were stuck on pub walls showing a cowering Hitler with the slogan 'Stamp on the Blighter'. Colossal amounts were raised, even though drinkers were already heavily funding the war effort through greatly increased duty on beer. Even teetotallers were moved to help. 'Aquarius' wrote to the *Western Mail*:

> While I am profoundly convinced that those who indulge in intoxicating drinks are grievously mistaken, I am nevertheless compelled to realise that they are shouldering a very heavy proportion of the financial burden imposed by the war. If only therefore as a thank you offering for the blessings of temperance, I feel that total abstainers should be only too pleased to contribute to a special fund for the purchase of Spitfires. I am enclosing £1 for the Spitfire Fund, and hope my fellow abstainers will follow suit.

The Custom House on Bute Street is the centre of attention after being hit during the Second World War. The Quebec, behind, became famous in the 1960s for its live music, notably from guitarist Vic Parker.

The Second World War cemented the pub in the public's affections, and in 1961 the eighty-year drought on a Sunday came to an end when a new Licensing Act allowed people in Wales to vote to repeal Sunday closing. But many local houses were not able to benefit from the new hours for long. For having survived enemy bombing during the war, real destruction followed when the council decided to redevelop the city. This was bad news for the small community pubs tucked away in the streets of back-to-back houses. When the houses went, so did the pubs. In Canton a whole host of street-corner locals were lost when the Wellington Street area was redeveloped in the early 1970s, such as the Rover Vaults, the Greyhound, the Red Cow and the Duke of York.

In the city centre, the pub was in peril from major shopping developments. Queen Street, which in 1903 had seven pubs and hotels, was steamrollered by the big stores. Popular pubs like the Tivoli, demolished in 1960 to make way for the Queen's Court development, and the Taff Vale, closed in 1977 for the St David's Centre, vanished. Both once had busy upstairs lounges as well as ground-floor bars. Now shoppers had to go thirsty. Where the temperance campaigners had failed, the retail giants had succeeded in creating a dry desert.

With the sharp decline of the coal trade, the docks had fallen into deep depression and much of the housing and many of the pubs were swept away in the early 1960s. Tiger Bay no longer roared, it just wept. Whole communities were scattered to distant housing estates. Newtown, a tightly packed bundle of streets just off Bute Street, once known as 'Little Ireland', was razed to the ground in 1967 along with pubs like the Crichton Arms, the Cambridge and the East Dock, all in Tyndall Street. All that was left were memories, particularly of one of Cardiff's most memorable men – 'Peerless' Jim Driscoll. He was not only one of the Britain's best boxers, but he also linked up pubs across Cardiff, having run one in Newtown, trained in others in the docks and been long remembered in a third across town, which became a shrine to his name.

The Duke of York in Wellington Street waits for its delivery from Brain's dray, parked opposite, in 1971. Within four years it had been demolished.

The Taff Vale on Queen Street was shut in 1977.

Jim Driscoll was born in Ellen Street, Newtown in 1880 and died little more than forty-two years later, just a few doors away in the Duke of Edinburgh pub in the same row of terraced houses. But during his short life his reputation soared out of the street and he became British and European featherweight champion, and a celebrity not just in Cardiff, Wales and Britain but also in the United States, where he beat the world champion in a non-title fight. Like many successful boxers, he took hard body blows early in life. His railwayman father died within months of his birth, run over by a train in the goods yard. His mother struggled to bring up a large family by bagging potatoes in the docks. Driscoll also soon learned to use his hands to make money. His stepfather Frank Franklin set up a gymnasium in the Cape Horn, where he was landlord. Many other docks pubs had boxing rings and Driscoll sparred in them all, or fought for a few shillings in the backyards of houses like the Blue Anchor. It was a brutal business. The ring at the North and South was known as the 'Blood Kitchen'.

Boxing as well as beer was once soaked into the pub life of Cardiff, and the men who made money out of it were tougher than the fighters. Bob Downey was known as Long John Silver as he only had one leg, a wooden crutch and a parrot. But no one dared call him that to his face. He had a monopoly on the sawdust trade in the docks, which was big business in the days when tons of it was used to mop up the blood and beer in the bars. He also ran the Bute Castle, where the parrot was trained to swear at policemen. Any trouble and Downey would vault the bar and crack heads with his crutch. He was the driving force behind the famous 'Club de Marl', the wasteland in the docks where crowds of drinkers would gather on Sunday with a few barrels of beer to get round the Sunday Closing Act. Downey could always be relied on to provide a little extra entertainment, like a bare-knuckle fight, with a bit of betting on the side. He became Driscoll's first trainer.

Above: The Royal Oak in Roath in 1969, with the gymnasium behind.

Right: Landlady Kitty Flynn shows actor Michael Hewson Jim Driscoll's Lonsdale Belt in the bar of the Royal Oak in 1988, when the Sherman Theatre in Cardiff was showing a play about the boxer's life.

Driscoll never escaped the ring. His fans wouldn't let him. He was still boxing at the highest level at forty. He also never escaped from the licensed trade. In 1907 he married Edie, the daughter of Bob Wiltshire, landlord of the Cambrian in St Mary Street. In 1918 he took over the Duke of Edinburgh back home in Ellen Street from his father-in-law. He had served in the Welsh Horse during the war and was known for tapping casks of beer while sitting astride the barrel. 'Army training,' he would grin. The small pub became a magnet for Cardiff's boxers and boxing fans. But after his early death from consumption in 1925 and the closure of the Duke, another house became a shrine to Peerless Jim.

Like the long-gone pubs in the docks, the Royal Oak in Roath has a boxing ring upstairs where champions have trained. Like some of them it also has an elaborate bar back. In addition, unlike the docks, it boasts fine stained-glass screens across the windows, featuring the heavyweight literary combination of Shakespeare, Byron and Chaucer. And unlike any other Brain's pub in town it still serves beer – Brain's SA – direct from the barrel. From 1947 this showcase pub on Broadway was run by Jim 'Ocker' Burns, who won three rugby caps for Wales, but it was his boxing connections which knocked out the locals. His father Tom was Jim Driscoll's uncle. When Jim Burns died in 1971, his daughter Kitty Flynn and her husband Mike took over. She had been born in the Duke of Edinburgh, and she further filled the bar with tributes to the ring legend, until she retired in 2003.

But her name is still prominent in the licensed trade, as Brain's renamed their Irish bar, Mulligan's, after her in 2002. It was an appropriate choice. Kitty Flynn's in St Mary Street had originally been the Cambrian, from where Jim Driscoll had eloped with the landlord's daughter. The only surprise is that there is no pub in Cardiff called Peerless Jim, though there is a statue of him at the top of Bute Street, gazing longingly across at the Golden Cross, hoping for one last beer and one final fight.

Mulligan's Irish bar in St Mary Street, which was named after Kitty Flynn in 2002.